T0184697

Wissenschaftliche Reihe Fahrzeugtechnik Universität Stuttgart

Reihe herausgegeben von
Michael Bargende, Stuttgart, Deutschland
Hans-Christian Reuss, Stuttgart, Deutschland
Jochen Wiedemann, Stuttgart, Deutschland

Das Institut für Verbrennungsmotoren und Kraftfahrwesen (IVK) an der Universität Stuttgart erforscht, entwickelt, appliziert und erprobt, in enger Zusammenarbeit mit der Industrie, Elemente bzw. Technologien aus dem Bereich moderner Fahrzeugkonzepte. Das Institut gliedert sich in die drei Bereiche Kraftfahrwesen, Fahrzeugantriebe und Kraftfahrzeug-Mechatronik. Aufgabe dieser Bereiche ist die Ausarbeitung des Themengebietes im Prüfstandsbetrieb, in Theorie und Simulation. Schwerpunkte des Kraftfahrwesens sind hierbei die Aerodynamik, Akustik (NVH), Fahrdynamik und Fahrermodellierung, Leichtbau, Sicherheit, Kraftübertragung sowie Energie und Thermomanagement – auch in Verbindung mit hybriden und batterieelektrischen Fahrzeugkonzepten. Der Bereich Fahrzeugantriebe widmet sich den Themen Brennverfahrensentwicklung einschließlich Regelungs- und Steuerungskonzeptionen bei zugleich minimierten Emissionen, komplexe Abgasnachbehandlung, Aufladesysteme und -strategien, Hybridsysteme und Betriebsstrategien sowie mechanisch-akustischen Fragestellungen. Themen der Kraftfahrzeug-Mechatronik sind die Antriebsstrangregelung/Hybride, Elektromobilität, Bordnetz und Energiemanagement, Funktions- und Softwareentwicklung sowie Test und Diagnose. Die Erfüllung dieser Aufgaben wird prüfstandsseitig neben vielem anderen unterstützt durch 19 Motorenprüfstände, zwei Rollenprüfstände, einen 1:1-Fahrsimulator, einen Antriebsstrangprüfstand, einen Thermowindkanal sowie einen 1:1-Aeroakustikwindkanal. Die wissenschaftliche Reihe „Fahrzeugtechnik Universität Stuttgart" präsentiert über die am Institut entstandenen Promotionen die hervorragenden Arbeitsergebnisse der Forschungstätigkeiten am IVK.

Reihe herausgegeben von

Prof. Dr.-Ing. Michael Bargende
Lehrstuhl Fahrzeugantriebe
Institut für Verbrennungsmotoren und
Kraftfahrwesen, Universität Stuttgart
Stuttgart, Deutschland

Prof. Dr.-Ing. Hans-Christian Reuss
Lehrstuhl Kraftfahrzeugmechatronik
Institut für Verbrennungsmotoren und
Kraftfahrwesen, Universität Stuttgart
Stuttgart, Deutschland

Prof. Dr.-Ing. Jochen Wiedemann
Lehrstuhl Kraftfahrwesen
Institut für Verbrennungsmotoren und
Kraftfahrwesen, Universität Stuttgart
Stuttgart, Deutschland

Weitere Bände in der Reihe http://www.springer.com/series/13535

Omar Abu Mohareb

Efficiency Enhanced DC-DC Converter Using Dynamic Inductor Control

Omar Abu Mohareb
Chair of Automotive Mechatronics, IVK, Faculty 7
University of Stuttgart
Stuttgart, Germany

Dissertation University of Stuttgart, 2018

D93

ISSN 2567-0042   ISSN 2567-0352   (electronic)
Wissenschaftliche Reihe Fahrzeugtechnik Universität Stuttgart
ISBN 978-3-658-25146-8   ISBN 978-3-658-25147-5   (eBook)
https://doi.org/10.1007/978-3-658-25147-5

Library of Congress Control Number: 2018967681

This Springer Vieweg imprint is published by the registered company Springer Fachmedien Wiesbaden GmbH part of Springer Nature
The registered company address is: Abraham-Lincoln-Str. 46, 65189 Wiesbaden, Germany

Acknowledgement and Dedication

This work was created during my work as a research scientist at the Institute of Internal Combustion Engines and Automotive Engineering (IVK) at University of Stuttgart and the Research Institute of Automotive Engineering and Vehicle Engines Stuttgart (FKFS).

My thanks and appreciation to Prof. Dr.-Ing. Hans-Christian Reuss for persevering with me as my advisor throughout the time it took me to complete this research and write the dissertation.

I am very thankful to the member of my dissertation committee, Prof. Giorgio Rizzoni who has generously given his time and expertise to review and better my work. I must acknowledge as well Dr. Qays Noaman who inspired, assisted, advised, and supported my research and writing efforts over the years.

I am grateful to my colleagues at the IVK and FKFS for the pleasant working atmosphere and excellent cooperation, especially Dr.-Ing. Michael Grimm, Head of Department Automotive Mechatronics and Electronics, for his continuous support and encouragement.

I would like also to thank my parents, family and friends for their support and trust.

Above all, I dedicate this dissertation to my beloved wife Alia who encouraged and supported me to pursue my dreams and finish my dissertation and to my little lovely daughter Ghaida for her boundless love and patience she granted me during the preparation of this work.

Omar Abu Mohareb

Table of Contents

List of Figures

List of Tables

Abbreviations and Nomenclatures

Abbreviations

AC	*Alternating current*
ADC	*Analog to Digital Converter*
BBC	*Boost Battery Charger*
BJT	*Bipolar Junction Transistor*
CAN	*Controller Area Network*
CC	*Constant Current*
CCM	*Continuous Conduction Mode*
CV	*Constant Voltage*
DC	*Direct Current*
DCM	*Discontinuous Conduction Mode*
DIC	*Dynamic Inductor Control*
DPWM	*Digital Pulse Width Modulator*
ESR	*Equivalent Series Resistance*
EV	*Electric Vehicle*
FKFS	*Research Institute of Automotive Engineering and Vehicle Engines Stuttgart*
IVK	*Institute for Internal Combustion Engines and Automotive Engineering at University of Stuttgart*
LCD	*Liquid-Crystal Display*
Li-ion	*Lithium-ion*
LPF	*Low-Pass Filter*
MET	*Maximum Energy Transfer*
MODE I	*Transistor in the BBC is switched on*
MODE II	*Transistor in the BBC is switched off*
MOSFET	*Metal-Oxide-Semiconductor Field-Effect Transistor*
ODE	*Ordinary Differential Equation*
PCB	*Printed Circuit Board*
PI	*Proportional-Integral*
PWM	*Pulse Width Modulation*
RMS	*Root Mean Square*
SMPS	*Switch-Mode Power Supply*
SOC	*State of Charge*
SSA	*State-Space Averaging*

THD	*Total Harmonic Distortion*
ZVRT	*Zero Voltage Resonant Transition*

Nomenclatures

Symbol	Unit	Description
A_e	m^2	*Core effective cross-sectional area*
Ax	-	*Magneto-resistive current sensor number x*
B	T	*Flux density*
B_{lim}	-	*Limit factor of the initial control current*
B_{max}	T	*Maximum flux density*
C	F	*Capacitance*
C1	-	*BBC output capacitor*
D	-	*Duty cycle*
D1	-	*BBC diode*
D2	-	*DIC diode*
D_{min}	-	*Minimum duty cycle*
d_1	m	*Toroid core outer diameter*
d_2	m	*Toroid core inner diameter*
E_a	J	*Inductor energy delivered to load*
E_L	J	*Inductor energy at end of MODE I*
E_{max}	J	*Maximum transferred energy achieved by DIC*
E_r	J	*Transfer energy per cycle for keeping converter output values constant*
f	Hz	*Switching frequency*
G_{LPF_i}	-	*Low-pass filter transfer function of current sensor*
g	-	*Converter control switch*
H	A/m	*Magnetic field intensity*
H_1	A/m	*Magnetic field intensity at primary side*
H_2	A/m	*Magnetic field intensity at secondary side*
H_{cs}	-	*Current sensor transfer function*
H_{max}	A/m	*Maximum magnetic field intensity*
h	m	*Toroid core height*
I_1	A	*Maximum inductor current at end of MODE I*
I_2	A	*Minimum inductor current at end of MODE II*

I_{av}	A	Average input (inductor) current
I_{ctr}	A	Control current
I_{ctr_max}	A	Maximum control current
I_{ctro}	A	Initial control current
I_{ctr_r}	A	Control current reference value
I_{L_RMS}	A	RMS inductor current
I_o	A	Average output current
I_{o_ref}	A	Reference output current
i	A	Inductor current
i_c	A	Capacitor current
i_D	A	Diode current
i_o	A	Output (load) current
i_T	A	Transistor current
K	-	Iteration
L	H	Inductance
L_c	H	Inductance value of critically damped converter
L_{CCM}	H	Minimum inductance required for CCM operation
L_e	H	Effective inductance
L_{max}	H	Maximum inductance achieved by DIC
L_{MET}	H	Inductance value that achieves MET
l_e	m	Core effective mean magnetic path length
N	-	Coil turns ratio
N_1	-	Coil number of turns at primary side
N_2	-	Coil number of turns at secondary side
P1	-	BBC controller unit
P2	-	Dynamic inductor controller unit
P_{ctr}	W	DIC conduction losses
P_L	W	Inductor conduction losses
P_{loss}	W	DC-DC converter losses
P_{loss_all}	W	BBC and DIC conduction losses
PWM1	-	BBC control signal
PWM2	-	Dynamic inductor control signal
Q	-	Quality factor
Q1	-	BBC N-MOS transistor
Q2	-	DIC N-MOS transistor

R	Ω	*Load resistor*
R_b	Ω	*Battery internal resistor*
R_c	Ω	*Capacitor internal resistor*
R_D	Ω	*Diode internal resistor*
R_e	Ω	*Equivalent conduction losses resistor referred to converter secondary side*
R_{ee}	Ω	*Equivalent conduction losses resistor referred to converter primary side*
R_L	Ω	*Inductor internal resistor*
R_{on}	Ω	*transistor internal resistor*
R_s	Ω	*DC power source internal resistor*
R	-	*Ripple factor of inductor current*
S	*Hz*	*Complex frequency of Laplace Transform*
T	*s*	*Transistor switching period*
T_r	*s*	*Response time*
t	*s*	*time*
t_{on}	*s*	*Transistor on-time*
V_b	*V*	*Battery open-circuit voltage*
V_{b_max}	*V*	*Maximum allowed battery voltage*
V_D	*V*	*Diode voltage drop*
V_c	*V*	*Capacitor voltage*
V_{ctr}	*V*	*Dynamic inductor control input voltage*
V_e	*V*	*Equivalent conduction losses voltage referred to converter secondary side*
V_{eq}	*V*	*Equivalent source voltage referred to converter secondary side*
V_{in}	*V*	*Source voltage at the DC-DC converter input*
V_L	*V*	*Inductor voltage*
V_o	*V*	*Output (load) voltage*
V_{o_ref}	*V*	*Reference output voltage*
V_s	*V*	*Source voltage*
V_T	*V*	*BJT voltage drop voltage*
xC	*1/h*	*Battery charging or discharging at x c-rate relative to its maximum capacity*
+x	*A*	*Positive limiting factor for ΔI*

Greek Symbol	Unit	Description
$\mathcal{F}_1$	A	*Magnetomotive force at primary side*
$\mathcal{F}_2$	A	*Magnetomotive force at secondary side*
H	-	*Converter efficiency*
η_{all}	-	*System overall efficiency (BBC and DIC)*
η_{max}	-	*Maximum Converter efficiency achieved by DIC*
Δcff_1	%	*Efficiency throw with DIC disabled*
Δeff_2	%	*Efficiency throw with DIC enabled*
ΔI	A	*Difference between inductor's maximum and minimum current values*
ΔI_{ctr}	A	*Small control current perturbation*
ΔI_L	A	*Difference between average input current and maximum allowed control current*
ΔI_{min}	A	*Small perturbation of control current maximum limit*
Φ	Wb	*Magnetic flux*
Φ_1	Wb	*Magnetic flux at primary side*
Φ_2	Wb	*Magnetic flux at secondary side*
M	H/m	*Permeability*
μ_1	H/m	*Permeability caused by the input current only*
μ_{max}	H/m	*Maximum permeability*
$\mathbb{R}$	-	*System operating region*
τ	s	*Delay time*
τ_{ADC}	s	*Delay time of analog to digital converter*
$\tau_{\mu c}$	s	*Delay time of microcontroller*
τ_{PWM}	s	*Delay time of DPWM unit*
V	-	*Lyapunov function*
ω	rad/s	*Angular frequency*
ζ	-	*Damping ratio*

Abstract

Switched-mode DC-DC converters are preferably used when efficient voltage-level conversion is required. Conventional DC-DC converter designs suffer from efficiency and performance degradation at higher output power rates due to the increased conduction and root mean square losses. The main cause of these higher losses is the degradation in the inductance value when higher DC bias current is passing through the inductor.

The basic buck and boost DC-DC converter topologies are remarkably influenced by the degradation in the inductance value, as they are designed and optimized around certain operating points. These conventional design methods usually overlook the change in the inductance value and suffer from higher losses at higher output power rates.

The work in this dissertation proposes a novel dynamic inductor control concept to overcome these drawbacks throughout a general method that can be flexibly applied to various DC-DC converter topologies. The proposed method controls the inductance value at all operating points and handles the physical limits of the inductor during operation, especially, at extreme duty cycle ratios and higher output power rates. Thus, the DC-DC converter can efficiently operate again around its physically optimized conditions and ease balancing DC-DC converter circuit performance, cost and size.

The state-space averaging method is implemented to provide steady-state and small-signal models of the converter using a set of second order linear differential equations to accurately evaluate the behavior of the non-ideal and non-linear DC-DC converter with the proposed control strategy. The detailed general state-space averaging model for non-ideal DC-DC converter with battery load is obtained to show the dissipative elements effects and the battery load influence on the model and its behavior.

Moreover, the principle of maximum energy transfer based on dynamic inductor control concept is introduced. The principle investigates the limits of the inductance value at which high efficiency is achieved without losing converter's dynamic response against load changes.

The new proposed dynamic inductor control concept is simulative evaluated in two basic DC-DC converter topologies and experimentally demonstrated in a prototype boost DC-DC converter with battery load. The simulation results have a good agreement with the experimental results and both results have shown outstanding potential to enhance the efficiency compared to the state-of-the-art converters at various operating conditions.

The improved converter efficiency improves system temperature and reliability. This may also help reducing the physical package sizes as well as the cooling required for the converter, and thus, reducing the total system cost and weight.

The dynamic inductor control can dramatically help improving the DC-DC converter efficiency in applications that have variable load, such as battery chargers and converters used to feed the low-voltage instruments from the traction battery in electric vehicles. The dynamic inductor control has also proved its effectiveness in improving DC-DC converter efficiency used with applications that have variable source voltage, such as the converters used with renewable energy sources. This makes the proposed novel dynamic inductor control generally applicable to various DC-DC converter types used in many applications where the inductance value needs to be optimized at different loads or source voltages and does not limit its scope to the basic buck and boost DC-DC converter topologies.

Kurzfassung

Bei effizienter Gleichspannungswandlung werden vorzugsweise getaktete DC-DC-Konverter eingesetzt. Herkömmliche DC-DC Konverter verlieren an Effizienz und Leistung bei höheren Ausgangsleistungen aufgrund von Leitungs- und „Root-Mean-Square"-Verlusten. Das Hauptproblem für dieses Verhalten ist die höhere Gleichstrombelastung der Spule und somit die Verringerung der Induktivität.

Die „Buck-" und „Boost"-Konverter sind im Wesentlichen von dieser Induktivitätsverringerung beeinflusst, sobald sie von ihrem ausgelegten und optimierten Betriebspunkt abweichen. Bei diesen Spannungswandlern wird die Veränderung der Induktivität nicht berücksichtigt und somit steigen die Verluste bei höherer Belastung.

In dieser Dissertation wird ein neuartiges Konzept mit einer dynamischen und variablen Induktion erarbeitet, um die oben genannten Nachteile für verschiedene DC-DC-Konverter zu vermeiden. Diese Methode regelt die Induktivität für jeden Betriebspunkt und berücksichtigt die physikalischen Grenzen der Induktivität, vor Allem im Grenzbereich, bei hohen „Duty Cycle" und hoher Last. Der Spannungswandler kann dadurch effizienter in einem Bereich um seinen physikalisch optimierten Betriebspunkt arbeiten und kann damit kosten- und bauraumoptimiert ausgelegt werden.

Die Zustandsraumdarstellung ermöglicht eine Bestimmung des Gleichgewichtszustands und Kleinsignal-Modells des Konverters. Anhand dieser Modelle lässt sich das Verhalten mit linearen Differentialgleichungen zweiter Ordnung des nicht idealen und nicht linearen DC-DC-Konverters inklusive der Regelstrategie beschreiben. Ein detailliertes Zustandsraummodel für nicht ideale Wandler mit einem Akkumulator als Last wird verwendet um die Verlustanteile und den Einfluss eines Akkumulators sowie das Verhalten des Gesamtsystems zu beschreiben.

Nachdem das Prinzip der maximalen Energiewandlung basierend auf der dynamischen Reglung der Induktivität erläutert wurde, wird das Limit der Induktivität bei dem Maximum der Effizienz untersucht ohne die Dynamik des Systems gegen Laständerungen zu verringern.

Die neuartige dynamische Induktivitätsregelung wurde simulativ für zwei DC-DC-Konverter evaluiert und experimentell anhand eines Boost-Konverters mit einem Akkumulator als Last experimentell bestimmt. Die Ergebnisse der Simulation und die Messergebnisse vom Prototypen weisen eine Übereinstimmung auf und beide zeigen die Steigerung der Effizienz im Vergleich zu den aktuellen Spannungswandlern für unterschiedliche Betriebspunkte auf.

Die verbesserte Effizienz des Konverters verbessert ebenso die thermische Eigenschaft und die Zuverlässigkeit. Diese Verbesserungen bringen eine Reduzierung des Bauraums und des Kühlungsbedarfs mit sich und reduzieren somit die Gesamtkosten und das Gewicht.

Die dynamische Induktivitätsregelung kann bei der Verbesserung von DC-DC-Konvertern mit einer variablen Last helfen, wie zum Beispiel bei Batterieladegeräten oder der Versorgung von Niedervoltgeräten, die von der Traktionsbatterie versorgt werden. Diese Regelung der Induktivität kann auch für die Effizienzsteigerung bei Applikationen mit variabler Spannungsversorgung sowie regenerativen Energiequellen verwendet werden. Das macht dieses Konzept allgemein anwendbar in Applikationen mit variabler Last oder variabler Eingangsspannungen.

1 Introduction

The wide spread of battery-powered new technologies such as smart phones, tablets computers and electric vehicles (EV) are forcing the development of more efficient and powerful battery chargers. The key technology to satisfy these requirements are the switched-mode DC-DC converters, which are used to "pump" the charging current into the battery.

The higher conduction and root mean square (RMS) losses of some DC-DC converter topologies at higher output power rates are a significant problem in charging applications [1]. These losses are mainly generated from the non-ideal storage and switching elements of the converter. A part of these losses is directly related to the inductor and they are dependent on its value. The growing conscious of energy conservation and environmental protection are forcing the development of more efficient DC-DC converter technologies. With this scope, the DC-DC converter must be designed to ensure the converter's load or the battery can always be efficiently charged at all operating points.

1.1 Motivation

The inductor is a major component in many switched-mode DC-DC converter topologies. The inductance value affects the DC-DC converter operating mode and its conduction losses. The conventional design methods of the DC-DC converters usually overlook the change in the inductance value, and thus the RMS ripple, due to the change of the DC bias current passing through it at different operating points. The problem is becoming worse in higher charging current applications where the inductance value is dropping and the RMS losses are becoming higher causing the efficiency to drop considerably. Accordingly, inductance control is required to maintain the maximum possible inductance value at all operating points.

The inductance control is used to assure minimal conduction and RMS losses at all operation points of the converter and allow maximum energy transfer (MET) from the power source to the load.

© Springer Fachmedien Wiesbaden GmbH, part of Springer Nature 2019
O. Abu Mohareb, *Efficiency Enhanced DC-DC Converter Using Dynamic Inductor Control*, Wissenschaftliche Reihe Fahrzeugtechnik Universität Stuttgart, https://doi.org/10.1007/978-3-658-25147-5_1

The conduction and RMS losses issues are dramatically higher by low-voltage and high-current DC-DC converters, such as DC-DC boost converters, where the load impedance is a fraction of the ohm. The inductor series resistance and the capacitor equivalent series resistance (ESR) are so significant in the converter's properties and dynamic performance [2]. The state of the art in designing an efficient converter is to precisely analyze and model its behavior. Modeling and simulation of components and processes play an important role in developing DC-DC converter design. Such analysis can be used to implement the proper inductance controller to achieve the maximum possible efficiency at different operating points.

Knowing these points will help identifying the MET operating point between the source and the battery, and thus, the fastest and the most efficient possible charging conditions.

1.2   Scope

The work in this dissertation is conducted to evaluate the feasibility of employing an inductance control with a DC-DC boost converter to identify the possible MET operating point under different operating conditions.

The boost DC-DC converter is one of the basic switched-mode DC-DC converter topologies. It outputs a voltage that is greater than the input voltage and draws an input current through the inductor that is greater than the output current. This makes the DC-DC boost converters a good candidate for investigating inductance control. Therefore, the novel dynamic inductor control (DIC) is introduced in this dissertation and its effectiveness of improving the converter's efficiency is demonstrated with great details using the boost DC-DC converter.

This work aims to accurately analyze, model and simulate DC-DC boost converter's steady-state and transient behaviors in continuous conduction mode (CCM) to help implementing the MET concept and the dynamic inductor control. The state-space averaging (SSA) model and canonical model are implemented to provide accurate steady-state and small-signal models of the converter in order to describe the behavior of the non-linear boost converter.

The non-ideal models of the converter are introduced to clarify the effect of dissipative elements (e.g. internal resistances) on the converter's characteristics and performance. The theoretical results are verified by simulation and experimental results to show the accuracy of the models. The models are used to find the system's transfer function in order to design the proper controller. The proposed dynamic inductor control is simulated and then experimentally validated to prove its effectiveness.

It is worth to mention that the analysis methodology and the introduced inductance control strategy in this work is a general approach which can be similarly implemented to other switched-mode DC-DC converters, as shown in section 5.4 afterward.

1.3   Dissertation Outline

This dissertation consists of eight chapters, which are organized as follows.

Chapter 1 highlights the necessity of efficient DC-DC converters, then describes the motivation and the scope of the dissertation in improving the converter's efficiency throughout using the dynamic inductor control concept.

Chapter 2 introduces a literature review on efficient DC-DC converter designs and lists out their limitations in dealing with the variation and degradation in the inductance value during operation. The factors which are influencing the inductance value are also in-details described in this chapter. The state-of-the-art inductor control techniques in DC-DC converters are addressed here and their drawbacks are summarized.

The general boost battery charger (BBC) mathematical models are conducted in chapter 3. The state-space averaging and the canonical models are derived. The principle of maximum energy transfer is also presented in this chapter. The obtained models are used in later chapters to design and simulate DC-DC converters with dynamic inductor control.

The novel dynamic inductor control concept is introduced in chapter 4, where the inductance value effect on efficiency and variable inductor structure are demonstrated. The control methodology and stability of the proposed DIC is also investigated in this chapter.

The MATLAB Simulink models used to simulate, evaluate and validate the DIC using the previously obtained mathematical models are described in chapter 5. This chapter includes the controller simulation method, simulation assumptions and the evaluation of the DIC on the boost DC-DC converter steady-state and dynamic behavior across different operating conditions. This chapter also includes a section for evaluating DIC with other DC-DC converter types. The maximum energy transfer concept evaluation is presented here as well.

Chapter 6 presents a prototype DC-DC converter with dynamic inductor control used to evaluate and validate the proposed strategy.

Chapter 7 presents the experimental results for the dynamic inductor control implemented in the prototype DC-DC converter with the modified inductor construction to validate the proposed control strategy across different steady-state and transient tests.

Conclusions and future work are summarized in chapter 8.

2 Literature Review and State of the Art

DC-DC converters are commonly used in power-conserving applications, like battery-operated equipment to regulate to a lower voltage, boost an input voltage or invert it to create a negative voltage. Efficiency is an important DC-DC converter characteristic as it may achieve efficiencies greater than 95% under optimum conditions. However, this efficiency is limited by dissipative losses in the components.

The converter efficiency impacts the electrical and thermal losses in the system. It affects the system's operating conditions, temperatures and reliability, and has a direct effect on the physical package sizes as well as the cooling required for the converter. These factors contribute to the total system cost, weight and performance.

2.1 Literature Review on Efficient DC-DC Converters

The conduction losses in DC-DC converters are the major losses in high-power density and high-gain converters. Numerous techniques have been proposed to reduce the conduction and RMS losses and improve the efficiency of DC-DC converters.

The conventional boost converters cannot provide a high DC voltage ratio due to the losses associated with the inductor, filter capacitor, switch and output diode. A DC-DC converter design is proposed in [3] to avoid extreme duty cycle in the conventional boost converters when high step-up voltage gain is required. Therefore and in order to increase the conversion efficiency and voltage gain, a modified boost-forward-flyback converter to achieve high step-up voltage gain without an extremely high duty ratio is proposed in this work. The proposed converter is based on the coupled-inductor, clamped circuit and pumping capacitor and used to provide the DC-AC inverter in an electric vehicle with a 400 V DC from a 48 V Li-ion battery system.

© Springer Fachmedien Wiesbaden GmbH, part of Springer Nature 2019
O. Abu Mohareb, *Efficiency Enhanced DC-DC Converter Using Dynamic Inductor Control*, Wissenschaftliche Reihe Fahrzeugtechnik Universität Stuttgart, https://doi.org/10.1007/978-3-658-25147-5_2

A similar design concept is proposed in [4] with fewer components in the clamp circuit to recycle the energy stored in the leakage inductance and increase the efficiency without extreme duty cycle.

Although, the proposed designs in [3] and [4] achieve high efficiency for certain operating points, it is not generally applicable to any other converter types and designs. The considerable amount of the additional components required in both of the proposed DC-DC converters shows the drawbacks in cost, size and weight.

A further improvement in the boost-forward-flyback converter is proposed by Cai in [5]. A high efficiency and high voltage step-up gain DC-DC converter consists of a boost converter cell and an isolated single-ended PWM resonant converter cell, with their outputs connected in series to achieve high voltage step-up gain is proposed in this work. The proposed converter uses quasi-resonant technology to achieve zero current switching for the diode on the secondary of transformer and reduce the power switch turnoff losses. The modified design does not require passive clamp or snubber circuit since the leakage inductor and the secondary resonant capacitor constitute the resonant tank through which the input power is transferred to the load.

The resonant converter cell in the proposed converter is sensitive to any drifts in its components values, especially when the switching frequency has a constant single value. This requires inductors and capacitors with very low tolerance values, and this implies more costs to the converter and limits the operating range around the optimal operating point.

A design optimization for a soft-switching bidirectional DC-DC power converter is introduced in the dissertation from Zhang in [6]. The proposed converter in [6] uses soft-switching operation, which can be considered a zero-voltage resonant transition switching technology to achieve a high power density bidirectional converter design.

The author in [6] has in-details introduced the significant impact of inductor design on the system performances, such as the realization of complementary control zero voltage resonant transition (ZVRT) soft-switching, device switching loss, system volume, inductor power loss. It is necessary to optimize the inductance with all the design considerations. In order to realize ZVRT, the inductance should be greater than a certain boundary value and a theoretical derivation method to optimize this value is also introduced. The introduced optimization method for the inductance value does not consider the dependency of inductance value on the current passes through the inductor at different operating conditions. This implies to over-engineer the inductor to stay above a certain boundary value at all operating points.

Choi in [7] has proposed a high efficiency DC-DC converter for low DC renewable sources with an improved active-clamped DC-DC converter by using a dual active-clamping circuit and controlled by asymmetrical pulse width modulation (PWM) technique to achieve zero-current switching by all output diodes. Nevertheless, the proposed full-bridge dual active-clamping circuit has relatively more complex structure, more number of switches with associated driving circuits, in addition to a considerable amount of other additional components.

The work in [8] has discussed the optimization issues between conduction modes, switching frequencies, efficiency and the limitations of use of elements such as inductor in low power DC-DC converters. Design challenge and tradeoffs for optimum switching frequency and inductance value are also discussed. The work suggested quality factor inductor for efficient energy storage operation at lower switching frequency. The optimization assumes that the inductor quality factor is directly proportional to the applied frequency, where inductance value is always constant. Nevertheless, in practice, the properties of the inductor are varying at different operating conditions and it is not constant.

A circuit topology of soft-switching PWM DC-DC converter with a high frequency link is presented in [9]. It is composed of H full-bridge inverter with a series PWM power switch in DC busline and a parallel capacitive snubber between DC busline, a flat high frequency transformer with a center tapped configuration, a full-wave diode rectifier, DC reactor in series with the load.

In addition to the complex design, the bulky high frequency transformer, which used to achieve the high-gain between the input and output voltage levels, and the additional power switch DC busline are drawbacks of the proposed design. Also, the additional power switch DC busline increases the conduction and switching losses remarkably for switching frequency of high frequency inverter power stage less than 10 kHz compared with the conventional hard-switching inverter type DC-DC converter.

An interesting comparative analysis with respect to the ripple current reduction drawn from fuel cell stack is presented in [10]. The performance of different DC-DC converter topologies i.e. boost converter, multi device boost converter, multiphase interleaved boost converters and multi device multiphase interleaved boost converters are compared, including ripple current, size of passive components, power losses and system efficiency. The analysis shows that the boost converter conventional topology could be employed in high efficiency and high performance applications, where the converter efficiency is inversely proportional to the input ripple current. The need for the complex interleaved boost converters could be avoided using high inductance with low input ripple current values.

The works in [11] and [12] are examples of multi-stages high-gain converter that uses high efficiency step-up DC-DC converter to feed a cascaded DC-AC inverter with a higher DC-link input voltage. Therefore, a high frequency transformer is obligatory used to prevent high duty cycle ratios. Therefore, they have considerable amount of additional components required in both designs showing drawbacks in cost, size and weight.

Many DC-DC converter designs are based on preventing high duty cycle ratio during the operation, where the efficiency may remarkably drop, as in [3], [4] , [5], [6], [7], [8], [9], [10], [11] and [12]. The efficiency is improved for certain or limited operating range through choosing certain physical components under certain switching frequency. In other words, the DC-DC converter design is physically optimized for high efficiency at certain or limited operating points or range and cannot handle the physical limits of the converter's components during operation at extreme duty cycle ratios.

2.2   Inductor in DC-DC Converter

Typically, a DC-DC converter is a circuit that uses a power switch, an inductor, and a diode to transfer energy from input to output. The basic components of the DC-DC converter circuit can be rearranged to form a step-down (buck) converter, a step-up (boost) converter, or an inverter (flyback) [13]. The inductor in all the basic topologies is used to avoid power losses and guarantee high efficiency in DC-DC converters, as it stores the energy in its magnetic field and releases it with negligible loss.

The boost converter is one of the basic switched-mode DC-DC converters topologies. It has an output voltage that is greater than the input voltage and draws an input current through the inductor that is greater than the output current. The higher output voltage of the DC-DC boost (step-up) converter compared to its input voltage is built up by storing the energy in the inductor through many switching cycles. The inductor stored energy provides together with the input voltage source a higher voltage at the output. The converter's energy needs to be greater than the required output energy and all losses in the circuit.

The inductance value is determined based on the desired output voltage, maximum load current and switching frequency. Three parameters should be considered when selecting the inductor for a DC-DC converter: first, the saturation current of the inductor must be higher than the converter's required peak current in order to be able to supply the necessary output power. Second, the designer must consider the DC resistance of the inductor. Finally, the physical size of the inductor should be considered. In order to lower the DC resistance, an inductor in a larger package could be considered if the physical size of the inductor is not prohibitive [14].

Selecting the right power inductor is a critical aspect of the DC-DC converter efficiency. Cost, size, resistance, current capability and efficiency drive the choice of inductor for most DC-DC switching converters, where many such applications specify the inductor value shown in the switching converter's data sheet or evaluation kit, but those values are usually specific to application or performance criteria [15]. Therefore, good understanding of inductor behavior is in the first place required.

The inductor in the DC-DC converters does not have constant parameters at all operating points. For example, the inductor effective series resistance is frequency dependent where the resistance can dramatically increase at higher frequency [16]. The inductance value itself is also a function of the current passing through it [17, 18, 19].

The total inductor losses R_L consist of DC resistance losses and switching frequency dependent components, such magnetic hysteresis loss, eddy-current loss, skin-effect losses in the conductor, proximity effect losses and radiation losses [20]. This loss resistance is primarily responsible for defin-ing the quality of the inductor [16, 20].

The mathematical determination of total inductor losses is impractical. Therefore, inductors are usually measured over the entire frequency range to determine its quality factor Q in Eq. 2.1 [16, 20]. The quality factor of an inductor is defined as the ratio of reactance ωL to the total inductor resistance R_L of an induction coil [20].

$$Q = \frac{\omega L}{R_L(f)}$$

Eq. 2.1

Therefore, this current chapter investigates the effect of the inductor's dynamic behavior and the effect of inductance value on the energy transferred from the converter's input to its output and its efficiency for a selected inductor with a certain number of turns N, core material and switching frequency f.

2.3   Effect of Inductor Current on Inductance Value

The inductor in the DC-DC converters does not have a constant inductance value at all operating points. It is sensitive to the input average current I_{av} passing through it [17, 18, 19]. Consider the toroid core inductor shown in Figure 2.1 with a typical inductor hysteresis loop shown in Figure 2.2, the inductance L could be approximately found according to Eq. 2.2, where, μ, A_e and l_e are the core permeability, effective cross-sectional area and effective mean magnetic path length, respectively [21, 22].

$$L = \frac{\mu\, N_1^2\, A_e}{l_e}$$

Eq. 2.2

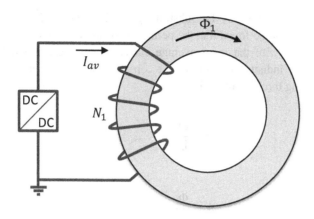

Figure 2.1: Toroid core inductor

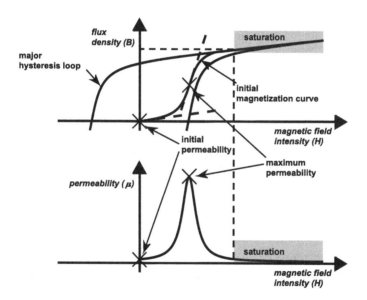

Figure 2.2:   Typical inductor hysteresis loop

If an input average current I_{av} is passing through the inductor and producing a flux Φ, the magnetic field intensity H and the flux density B are produced according to Eq. 2.3 and Eq. 2.4, respectively [21, 22]. The permeability describes the relationship between the flux density and the magnetic field intensity as in Eq. 2.5 [23]. By substituting Eq. 2.3 and Eq. 2.4 into Eq. 2.5 and Eq. 2.2, it can be found that the permeability and inductance are inversely proportional with the input average current as in Eq. 2.6 and Eq. 2.7. Therefore, the inductance value is not constant and it is a function of the current passing throughout the inductor.

$$H = \frac{N_1 \, I_{av}}{l_e} \qquad\qquad \text{Eq. 2.3}$$

$$B = \frac{\Phi}{A_e} \qquad\qquad \text{Eq. 2.4}$$

$$\mu = \frac{B}{H} \qquad\qquad \text{Eq. 2.5}$$

$$\mu = \frac{\Phi \, l_e}{A_e \, N_1 \, I_{av}} \qquad\qquad \text{Eq. 2.6}$$

$$L = \frac{\Phi \, N_1}{I_{av}} \qquad\qquad \text{Eq. 2.7}$$

2.4   Inductance Value Effect on Root Mean Square Losses

The amount of energy transferred by a real inductor in the DC-DC converter depends mainly on the inductance value L and the equivalent conduction losses resistance R_{ee}.

An inductor of inductance L connected to a DC source voltage V_s with an increasing DC current flowing through it from I_2 to final value of I_1, as shown in Figure 2.3, has a stored energy E_L of that in Eq. 2.8. The final value of the current I_1 and the stored energy are affected by the equivalent conduction losses resistance R_{ee} which limits the maximum current passing through the inductor.

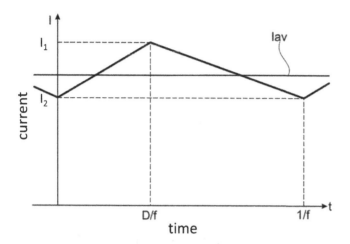

Figure 2.3: DC-DC converter inductor current

$$E_L = \frac{1}{2} L I_1^2$$
Eq. 2.8

When the inductor is connected to the load, it starts delivering its previously stored energy to it. Part of the inductor energy is dissipated as conduction losses on R_{ee}, which is the equivalent conduction losses resistance defined in Table 3.1. The inductor conduction losses for inductor's current shown in Figure 2.3 is given by Eq. 2.9 [24]. The I_{L_RMS} and I_{av} are the RMS and average inductor current and they are defined according to Eq. 2.10 and Eq. 2.11, respectively. The average inductor current and ripple factor r defined in Eq. 2.12 depend on the inductance value L [25] where D is the duty cycle which is the ratio between the transistor on-time t_{on} to the total switching period $T = 1/f$.

$$P_L = I_{L\_RMS}^2\, R_{ee} \qquad\qquad \text{Eq. 2.9}$$

$$I_{L\_RMS} = I_{av} \sqrt{1 + \frac{r^2}{12}} \qquad\qquad \text{Eq. 2.10}$$

$$I_{av} = \frac{I_1 + I_2}{2} \qquad\qquad \text{Eq. 2.11}$$

$$r = \frac{I_1 - I_2}{I_{av}} = \frac{\Delta I}{I_{av}} = \frac{D\,V_s}{I_{av}\,L\,f} \qquad\qquad \text{Eq. 2.12}$$

2.5   State-of-the-Art of Inductor Control in DC-DC Converters

In practice, properties of the physical components, like internal resistance, inductance and capacitance are drifting from its optimum value when the operating point is different set or because of temperature. The original high efficiency could be restored when the physical parameters are set back to their original values during all operating points.

The inductor is a major component that affects the efficiency, as described afterward in section 4.1. All the previously mentioned works in section 2.1 have not treated the variation in the inductance value while operating and assume that the inductance value is constant and not changing under different operating conditions, leading only to an improved efficiency over limited operating points.

Many works have investigated the variation in the inductance value and pursued controlling it using a magnetically controlled inductive device to overcome the changing inductance value at different operating points. The magnetically controlled inductive device is not new; the principles of the saturable-core control and magnetic amplifier were used in electrical machinery as early as 1888 although they were not identified as such [26].

In the US patent 7,378,828 B2 [27], an inductance control of a magnetic element for inhibiting saturation of the switched winding of the magnetic element is used to maintain a regulated output voltage. A magnetic element (inductor) having two orthogonal windings is used for rapidly compensating for load current variations in a low profile (light load) DC-DC converter. A negative magnetic feedback applied to magnetic element favorably alters the permeability function of magnetic element to extend the amount of load current the magnetic element may handle before reaching saturation. This allows the converter to remain at the boundary between discontinuous and continuous current mode (DCM-CCM) in order to improve its efficiency.

A magnetically controlled inductive device comprising an anisotropic material is in US patent US 7,256,678 B2 [28] introduced. It is used to control the magnetic flux conduction in a rolling direction by controlled domain displacement in the transverse direction. This invention is based on the possibility of altering the relative permeability of the core in relation to a primary magnetic field by altering the secondary (control) magnetic field which is at right angle to the primary. Different specially-designed core shapes are introduced and used where the primary and the secondary coils are winded in transverse direction. The transversal surface of the core is the cross-sectional area for the control flux density.

A magnetically controlled inductor in [29] is used to obtain pulse-width modulation control in a DC-DC converter by means of a magnetic amplifier. The magnetic amplifier has two control windings to control its saturation for the output voltage regulation and the over-current protection.

The invention in [30] is to provide a DC-DC converter circuit that operates over a wide range of input voltages and create a regulated output voltage by controlling the magnetic flux reset time of the core. The control circuit for the power converter provides control of the on-time and off-time of the power transformer by controlling the magnetic flux reset voltage produced on the secondary winding, thereby controlling the duty cycle of the power transformer, and thereby regulating the output of the power transformer.

A similar work of a "DC/DC converter with magnetic flux density limits" introduces a power stage control logic which is configured to prevent the transformer core in a DC-DC forward converter from saturating during the power transfer phase by causing the power transfer phase to terminate when the flux density information indicates that the flux density has reached or exceeded a pre-determined threshold [31]. The power stage circuit being configured to operate in a power transfer phase during which power is transferred from the input to the output and a reset phase during which flux density in the core of the transformer is reduced [32].

The drawbacks in the prior art are summarized as follow:

▪ The prior arts are meant for inhibiting saturation in inductor and/or maintaining a regulated output voltage without considering maintaining the highest possible efficiency at the required output voltage/current, as in [27], [28], [29], [30], [31] and [32]. For example, the inductor control associated with the applications mentioned in [28], including the variable choke in DC-DC converters, is meant as a voltage regulator and overlooks the efficiency considerations.

▪ The work in [27] controls the inductance value by remaining at the discontinuous current mode-continuous current mode (DCM-CCM) boundary and improves the converter's efficiency at light loads. This control strategy leads the DC-DC converter to have considerable RMS losses and thus low efficiency at heavy loads.

▪ A specially-designed inductor or unconventional complex-shape cores are required for the inductor control, where an inductor having two orthogonal windings in [27] and complex-shape cores in [28] are re-quired to implement the control strategy.

▪ The control function between the primary and control flux densities depends on the core shape and geometry, as in [28].

A novel dynamic inductor control is introduced in this work to address these drawbacks and overcome them, as described in chapter 4, and it is a general method that can be flexibly applied to various DC-DC converter topologies. The proposed method controls the inductance value at all operating points and handles the physical limits of the inductor during operation at extreme duty cycle ratios. Thus, the DC-DC converter can efficiently operate again at its physically optimized conditions and ease balancing DC-DC converter circuit performance, cost and size.

3 Boost Battery Charger Modeling

Switched-mode power supply (SMPS) or switched-mode DC-DC converter is the basic element for the majority of electronic equipment and domestic appliances. The light weight, high efficiency and low cost of SMPS's compared to linear power supplies make them favorable choice for stationary and mobile applications. The advantages of SMPS are marred by the non-linearity and dynamics of the system due to its sequential switching nature. So, a representative mathematical description for the SMPS boost converter; which provides a deep understanding of the operation of the switching converter and an easy-to-use accurate model, should be obtained.

There are two basic types of DC-DC SMPS: forward-mode converter and flyback-mode converter. Buck and Boost converters are the most elementary forward-mode and flyback-mode converters, respectively [1, 33]. DC-DC boost converter is one of the SMPS's basic topologies which can produce a higher regulated output voltage from a lower unregulated input voltage.

In this work, the DC-DC boost converter in continuous conduction mode (CCM), shown in Figure 3.1 is chosen to demonstrate the maximum energy transfer (MET) concept due to the following:

- Boost converter is generally less efficient than buck converter, since its inductor current flows to ground during the on-time and only a fraction during the off-time flows to the output [34]. The MET concept can be used to improve the boost converter efficiency.

- The peak and RMS inductor currents are larger in the discontinuous condition mode (DCM) boost converter. This implies that the boost converter should always be operated in the CCM for higher efficiency [35].

- CCM boost converters require higher inductance value to prevent the converter from falling into the DCM [25]. This requirement copes well with MET concept that forces the inductor to operate at its maximum possible inductance value.

- The boost converter is ideal choice for power factor correction circuits because it draws the input current continuously. Accordingly, the boost converter has much lower total harmonic distortion (THD) factor than

© Springer Fachmedien Wiesbaden GmbH, part of Springer Nature 2019
O. Abu Mohareb, *Efficiency Enhanced DC-DC Converter Using Dynamic Inductor Control*, Wissenschaftliche Reihe Fahrzeugtechnik Universität Stuttgart, https://doi.org/10.1007/978-3-658-25147-5_3

the buck converter and can be cascaded with other SMPS topologies [36, 37]. More efficient power factor correction circuit will assure more efficient converter with less THD factor.

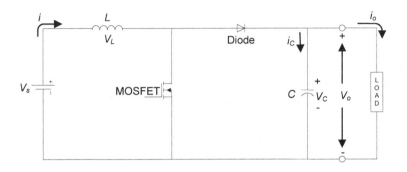

Figure 3.1:   Basic scheme of a boost battery charger

The boost converter has been treated in great detail because boost action appears in many converter combinations, like the isolated and multiple higher or lower output voltage converters [1]. A detailed model with the associated inductor, transistor, diode and capacitor losses, and capable of simulating the real performance of the converter is required in order to evaluate the forthcoming inductance control concept.

For high efficiency, the converter applies high-frequency transistor switch to pulse width modulate the input voltage into an inductor. The boost converter has two conduction modes: the continuous conduction mode (CCM) and the discontinuous conduction mode (DCM). In the CCM, the converter made to function in such that the inductor current is always positive. While in the DCM, the inductor current becomes zero before the end of the switching cycle. Accordingly, they represent different circuit topologies within each switching cycle. The boost converter mainly represents two circuit topologies in CCM and three circuit topologies in DCM [38, 39]. This dissertation investigates only the boost converter in the CCM, where DCM is beyond the scope of this work as it contradicts with the MET concept.

Three BBC mathematical models are developed to aid the design and analysis process requirements. The BBC has two main conduction modes depending on the switches status, namely the sequential conduction mode and the simultaneous conduction mode. In simultaneous conduction mode, both of the switches are both conducting at the same time. This mode occurs when the MOSFET is switching on or off and the diode is switching off or on; respectively. Modeling BBC in simultaneous conduction mode is not considered due to the negligible effect of the switching losses compared to the conduction losses at the decided switching frequency [40, 41, 42]. The BBC during the sequential conduction mode has two topologies and they are considered in details in subsection 3.2.1.

3.1 Modeling Non-Ideal Components

The ideal boost converter model is based on ideal elements of zero voltage drops and no series resistances. The simplified ideal model does not reflect the real performance or the transient behavior in reality; where the losses associated with the inductor, transistor, diode and capacitor affect the converter's characteristics; like, conversion ratio and efficiency [43].

In general, the boost converter system consists of a DC power source, storage elements with switching devices and a resistive or Li-ion battery load. The real non-ideal DC-DC boost converter contains non-ideal storage and switching elements. For this reason, it is essential to have a boost converter model that represents the physical response under the presence of those non-ideal elements.

The main losses in components are modeled in a form of resistors and voltage drops which can be directly extracted from the datasheets. Modeling the non-ideal Li-ion battery, DC power source, inductor, capacitor, MOSFET and diode for the non-ideal BBC is described in a detailed manner and for each component in the following subsections.

3.1.1   Lithium-Ion Battery Model

Many approaches can be found in the literature to model the Li-ion battery in a form of equivalent electronic networks; such as, the equivalent electronic network of linear passive resistances, capacitances and inductances [44] or the equivalent electronic network that consists of linear passive elements in combination with voltage source [45]. The main advantage of these models is that they can be analyzed and simulated together with the surrounding electronic circuits, such as the charger, using conventional electronic circuit analysis tools and simulators [46].

A battery model can be represented as in Figure 3.2 [47, 48]. This model is able to represent the battery voltage transient behavior. The RC networks parameters can be obtained from charging or discharging experiments by analyzing the voltage behavior as system response to current impulses [47].

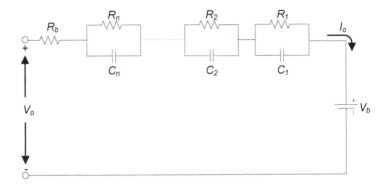

Figure 3.2:   Equivalent electronic networks for Li-ion battery

A more simplified battery model shown in Figure 3.3 is used in this work, where the transient behavior of the Li-ion battery voltage is out of the scope of this work. This simplified model of a voltage source V_b with series internal resistance R_b is valid under high frequency operation of more than hundreds of hertz [49], as in the DC-DC boost converter which typically operates in kilo-hertz range. It is worth to mention although the values of V_b and R_b are varying based on the battery state of charge (SOC), they can be considered constant values during many switching cycles of the boost converter [47, 49].

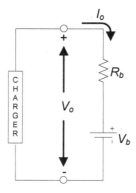

Figure 3.3:   Equivalent scheme of a simple Li-ion battery model

3.1.2   DC Power Source Model

The concept of ideal voltage source and internal resistance applies to all kinds of electrical sources and is useful for analyzing many types of electrical circuits. A practical electrical power source could be, according to Thévenin's theorem, represented as an ideal voltage source in series with internal resistance under DC operating conditions [50].

The equivalent circuit of a DC power source is shown in Figure 3.4. The power source's internal resistance R_s parameter can be determined by measuring the open-circuit voltage V_s and loaded voltage V_{in} for a defined current load. The R_s can be calculated as in Eq. 3.1.

$$R_s = \frac{V_s - V_{in}}{I} \qquad \text{Eq. 3.1}$$

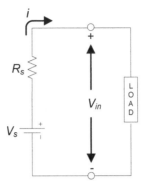

Figure 3.4:   DC power source modeled with one voltage source and internal resistance

3.1.3   Inductor Model

Real inductors exhibit two types of losses in form of core and winding losses. Both types are frequency dependent [23]. The core losses are in a form of hysteresis and eddy current losses and the copper losses are originating in the resistance of the wire, including skin and proximity effects [1, 38, 51]. A lumped parameter model for a ferrite core inductor, including core and winding losses is shown in [52] with an equivalent series resistance and an equivalent series inductance as a function of frequency.

At low frequencies below 20 kHz, the core losses in ferrite core inductors are negligible compared to the winding losses and the winding losses are mainly originating in the DC resistance of the wire [51, 53]. This resistance limits the converter's voltage conversion ratio. A suitable model that describes the inductor copper loss is shown in Figure 3.5, in which a resistor R_L is placed in series with the inductor. The resistance value is usually given by the manufacturer in the inductor's datasheet at certain operating points.

Figure 3.5:   Equivalent circuit of a non-ideal inductor

3.1.4 Capacitor Model

The filter capacitor in the DC-DC SMPS dissipates some energy on what is known as the equivalent series resistance (ESR) of a capacitor [54, 55, 56, 57]. The ESR of the capacitor is responsible for the energy dissipated as heat and it is defined as that portion of the capacitor's impedance that is responsible for the overall real power loss in the capacitor [54]. The ESR is also frequency dependent [54] and can be considered constant for a certain switching frequency, as in [55], [56] and [57]. Non-ideal capacitor equivalent circuit can be presented as an ideal capacitor with its series ESR R_c, as shown in Figure 3.6. In DC-DC converters, the ESR plays a major role in the converter's performance, as will be shown in the analysis in this chapter. The ESR is given in the capacitor's datasheet.

Figure 3.6:   Equivalent circuit of a non-ideal capacitor

3.1.5 Power Switches Models

The forward voltage of a MOSFET (BJT) can be modeled with reasonable accuracy as an on-resistance R_{on} [36]. In the case of a diode, IGBT or Thyristor, a voltage source (V_T for transistor and V_D for diode) plus an on-resistance (R_{on} for transistor and R_D for diode) yield a model of good accuracy [38]. Although the MOSFET can be modeled with the R_{on} only and the diode with the forward voltage source V_F only (R_D may be omitted if the converter is being modeled at a single operating point), both of them are modeled using voltage source and on-resistance in order to obtain a general purpose boost converter model, as shown in Figure 3.7 and Figure 3.8, respectively. It is worth to mention that the switching losses in the transistor and the diode are beyond the scope of this work and, therefore, they are not included in the model.

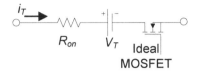

Figure 3.7:   Equivalent circuit of a non-ideal MOSFET

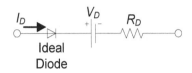

Figure 3.8:   Equivalent circuit of a non-ideal diode

By replacing the ideal components in the basic schematic of a boost battery charger in Figure 3.1 with their non-ideal models of the converter's components, a non-ideal scheme of a boost battery charger is obtained, as shown in Figure 3.9. Henceforth, this scheme is used to derive the different BBC models.

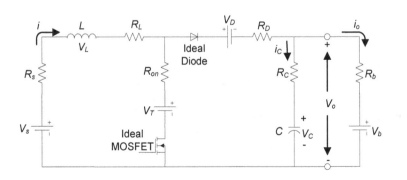

Figure 3.9:   Non-ideal scheme of boost battery charger

3.2 Boost Battery Charger Models

Switched systems such as DC-DC boost converters are a challenge to analyze since its model depends on whether a switch is opened or closed. In literature, there are various modeling methods. One approach is to derive the differential equations that describe the inductor current and capacitor voltage and then solve them in accordance with the boundary conditions of the switching cycle. Although the high accuracy of this method, it produces set of equations that require extensive computation [25].

In 1970s, Middlebrook published the models for DC-DC converters used today [58]. The approach is based on the assumption that the output voltage is constant during one switching cycle. This approach simplifies the mathematics and reduces the two systems to one. Although the simplified approach neglects the dynamic behavior of the circuit, it leads to insightful design equations which help defining the steady-state quiescent operating points. Hence, a small-signal linearized dynamic model with first-order AC terms around these quiescent operating points can be obtained and the converter transfer functions can be easily extracted.

The assumption of constant output voltage does make sense, where the output voltage switching ripple should be small in any well-designed converter. This assumption is called small-ripple approximation [38]. The neglect of the dynamic behavior provides the converter designer with a starting place. Once the inductance and capacitance values are determined from the obtained equations in CCM, those circuit parameters are used to observe the transient behavior. To obtain a more comprehensive and systematic method, Middlebrook averaged the circuit configurations for each switch state in a technique called state-space averaging (SSA).

The SSA method was the first approach to result in a complete linear model [58], which provides a deep understanding of the operation of the switching converters and an easy-to-use accurate model. The general steps of the SSA analysis are:

- find the exact characteristic differential equations at each converter's topology,

- generate the corresponding time-continuous non-linear differential equations,

- linearize the characteristic equations around the quiescent operating points and

- generate the transfer functions.

The ability of placing the resultant equations from the analysis in a standard matrix makes the SSA suitable for simulation programs, such as MATLAB. It also facilitates a suitable controller design for the converter through the simulation environment by accurately defining the open-loop and closed-loop converter's transfer functions.

3.2.1 State-Space Averaging Model

A detailed general SSA model for a non-ideal BBC with battery load is made in the following analysis to show the dissipative elements and the battery load effects on the model and its behavior. The notations used in this analysis are those commonly found in the literature. The capital letters denote the DC components of variables, while the small letters with a circumflex above them refer to their AC small-signal variations. The dual pointing angle is used to represent variables average values during one switching period.

The model purpose is to obtain a general accurate DC-DC boost converter model with a resistive R_b or battery load based on the scheme in Figure 3.9. The boost converter scheme with a resistive load is a special case from its counterpart with a battery load by setting the battery voltage to zero ($V_b = 0$) and substituting the resistive load value into the battery internal resistance ($R_b = R$). Accordingly, the general boost battery charger (BBC) scheme shown in Figure 3.9 can represent both cases.

The general BBC scheme shown in Figure 3.9 has two different topologies or modes in the CCM. In the first mode or MODE I shown in Figure 3.10 where the transistor is switched on, the inductor voltage $v_L(t)$ is given by Eq. 3.2. In contrary to the ideal boost converter model, the capacitor voltage $v_c(t)$ is not equal to the output voltage $v_o(t)$, due to the presence of capacitor internal resistance R_c. Two equations for the capacitor current $i_c(t)$ and the output voltage $v_o(t)$ are given by Eq. 3.3 and Eq. 3.4, respectively. By solving equations Eq. 3.3 and Eq. 3.4, the capacitor current $i_c(t)$ is given by Eq. 3.5.

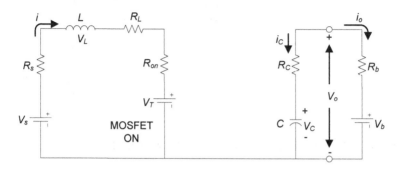

Figure 3.10: Non-ideal scheme of a BBC in MODE I

$$L\frac{di}{dt} = V_s - V_T - R_1 i(t)$$

Eq. 3.2

$$\text{where } R_1 = R_s + R_L + R_{on}$$

$$i_c(t) = C\frac{dv_c}{dt} = \frac{-(v_o(t) - V_b)}{R_b}$$

Eq. 3.3

$$v_o(t) = v_c(t) + R_c i_c(t) = v_c(t) + R_c\left(\frac{V_b - v_o(t)}{R_b}\right)$$

Eq. 3.4

$$i_c(t) = C\frac{dv_c}{dt} = \frac{-(v_c(t) - V_b)}{R_b + R_c}$$

Eq. 3.5

Similarly for the second mode or MODE II shown in Figure 3.11 where the transistor is switched off, the inductor voltage and the capacitor current are given in Eq. 3.6 and Eq. 3.7, respectively. The output voltage $v_o(t)$ in MODE II is defined in Eq. 3.8.

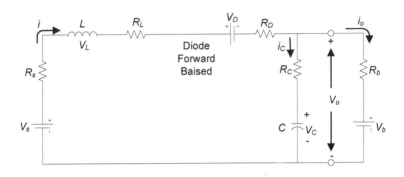

Figure 3.11:   Non-ideal scheme of a BBC in MODE II

$$L\frac{di}{dt} = V_s - V_D - v_o(t) - R_2 i(t)$$

Eq. 3.6

$$\text{where } R_2 = R_s + R_L + R_D$$

$$i_c(t) = C\frac{dv_c}{dt} = i(t) - i_o(t) = i(t) - \frac{(v_o(t) - V_b)}{R_b}$$

Eq. 3.7

$$v_o(t) = v_c(t) + R_c i_c(t) = v_c(t) + R_c\left(i(t) - \frac{(v_o(t) - V_b)}{R_b}\right)$$

Eq. 3.8

Solving Eq. 3.6, Eq. 3.7 and Eq. 3.8 for $v_L(t)$ and $i_c(t)$ yields to have inductor voltage and the capacitor current in terms of state variable $i(t)$ and $v_c(t)$, as depicted in Eq. 3.9 and Eq. 3.10, respectively. Applying the principles of inductor volt-second balance and capacitor charge balance, the DC components or average values of the inductor voltage and the capacitor current should equal zero, as in Eq. 3.11 and Eq. 3.12. Accordingly, the principles of inductor volt-second balance and capacitor charge balance for the BBC are as stated in Eq. 3.13 and Eq. 3.14, respectively. The equivalent circuits corresponding to these two equations can be constructed as depicted in Figure 3.12. The scheme in Figure 3.12 represents the BBC in steady-state, therefor the inductor and the capacitor can be removed from the scheme and they are drawn with dashed lines to indicate the steady-state analysis. The voltage and current dependent sources in Figure 3.12 can be replaced by a DC transformer with a conversion ratio of $((1-D)R_b/(R_b+R_c))\!:\!1$, as shown in Figure 3.13.

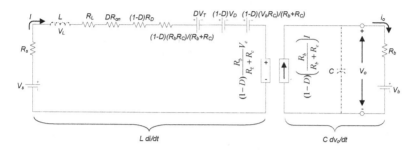

Figure 3.12: Non-ideal BBC equivalent circuits according to Eq. 3.13 and Eq. 3.14

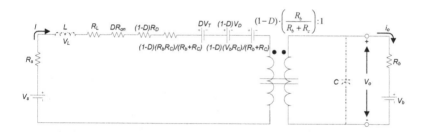

Figure 3.13: Non-ideal BBC equivalent circuit with DC transformer

$$L\frac{di}{dt} = V_s - V_D - \frac{R_c}{R_b+R_c}V_b - \frac{R_b}{R_b+R_c}v_c(t)$$
$$- \left(R_2 + \frac{R_bR_c}{R_b+R_c}\right)i(t)$$

Eq. 3.9

$$C\frac{dv_c}{dt} = \frac{R_b i(t) - v_c(t) + V_b}{R_b+R_c}$$

Eq. 3.10

$$D \times L \frac{di}{dt}\bigg|_{\text{MODE I}} + (1 - D) \times L \frac{di}{dt}\bigg|_{\text{MODE II}} = 0 \qquad \text{Eq. 3.11}$$

$$D \times C \frac{dv_c}{dt}\bigg|_{\text{MODE I}} + (1 - D) \times C \frac{dv_c}{dt}\bigg|_{\text{MODE II}} = 0 \qquad \text{Eq. 3.12}$$

$$\begin{aligned}
L \frac{di}{dt} = V_s - DV_T &- (1 - D)V_D - (1 - D)\frac{R_c}{R_b + R_c}V_b \\
&- (1 - D)\frac{R_b}{R_b + R_c}V_c - (R_s + R_L)I - DR_{on}I \qquad \text{Eq. 3.13} \\
&- (1 - D)R_D I - (1 - D)\frac{R_b R_c}{R_b + R_c}I = 0
\end{aligned}$$

$$C \frac{dv_c}{dt} = (1 - D)\frac{R_b}{R_b + R_c}I - \frac{V_c - V_b}{R_b + R_c} = 0 \qquad \text{Eq. 3.14}$$

Having the capacitor voltage equals to the output voltage in steady-state and by pushing all the elements from the primary side to the secondary side via the DC transformer, the steady-state DC equivalent circuit can be obtained as shown in Figure 3.14. This circuit describes the relationships between the input and the output voltages and currents as a function of the duty cycle and the equivalent conduction losses resistance R_e and equivalent conduction losses voltage V_e, as in Eq. 3.15 to Eq. 3.18.

$$V_{eq} = \left(\frac{1}{1 - D}\right)\left(\frac{R_b + R_c}{R_b}\right)V_s \qquad \text{Eq. 3.15}$$

$$I_o = (1 - D)\left(\frac{R_b}{R_b + R_c}\right)I \qquad \text{Eq. 3.16}$$

$$R_e = \left(\frac{1}{1-D}\right)^2 \left(\frac{R_b + R_c}{R_b}\right)^2 \left(R_s + R_L + DR_{on} + (1-D)R_D \right.$$
$$\left. + (1-D)\frac{R_b R_c}{R_b + R_c}\right)$$

Eq. 3.17

$$V_e = \left(\frac{1}{1-D}\right)\left(\frac{R_b + R_c}{R_b}\right)\left(DV_T + (1-D)V_D \right.$$
$$\left. + (1-D)\frac{R_c}{R_b + R_c}V_b\right)$$

Eq. 3.18

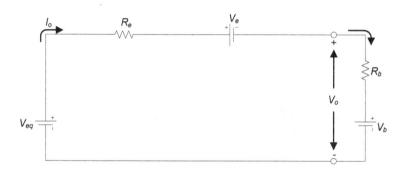

Figure 3.14: Non-ideal BBC steady-state DC equivalent circuit

From the steady-state DC equivalent circuit in Figure 3.14, the output voltage is given by Eq. 3.19 and Eq. 3.20. Solving these equations, the DC output voltage V_0, the DC output current I_0 and the DC input current I are derived in Eq. 3.21 to Eq. 3.23 as a function of duty cycle and internal resistances and voltage drops, where the BBC conversion ratio is limited by the internal resistances and voltage drops. The BBC efficiency can be obtained as in Eq. 3.24, where the DC source internal resistance R_s is considered a part of the converter losses. When R_e and V_e have non-zero values, the BBC efficiency is less than 100 %.

$$V_0 = V_{eq} - V_e - R_e I_0$$

Eq. 3.19

$$V_o = V_b + R_b I_o \qquad \text{Eq. 3.20}$$

$$V_o = \left(\frac{1}{1-D}\right)\left(\frac{R_b + R_c}{R_b + R_e}\right)V_s + \frac{R_e}{R_b + R_e}V_b - \frac{R_b}{R_b + R_e}V_e \qquad \text{Eq. 3.21}$$

$$I_o = \frac{\left(\frac{1}{1-D}\right)\left(\frac{R_b + R_c}{R_b}\right)V_s - V_b - V_e}{R_b + R_e} \qquad \text{Eq. 3.22}$$

$$I = \left(\left(\frac{1}{1-D}\right)\left(\frac{R_b + R_c}{R_b}\right)\right)\left(\frac{\left(\frac{1}{1-D}\right)\left(\frac{R_b + R_c}{R_b}\right)V_s - V_b - V_e}{R_b + R_e}\right) \qquad \text{Eq. 3.23}$$

$$\eta = \frac{V_o I_o}{V_s I} = \frac{\left(\frac{1}{1-D}\right)\left(\frac{R_b + R_c}{R_b + R_e}\right)V_s + \frac{R_e}{R_b + R_e}V_b - \frac{R_b}{R_b + R_e}V_e}{\left(\frac{1}{1-D}\right)\left(\frac{R_b + R_c}{R_b}\right)V_s} \qquad \text{Eq. 3.24}$$

Referencing to the two BBC topologies in Figure 3.10 and Figure 3.11 and using the control switch g in Eq. 3.25 as an additional input [25, 38, 58, 59], Eq. 3.13 and Eq. 3.14 can be rewritten in the standard state-space form where i and v_c are state variables and k, V_s, V_b and V_D are inputs, as in Eq. 3.26. This SSA model represents both the large-signal and linearized small-signal of the BBC.

$$g = \begin{cases} 1, & 0 \le t < t_{on} \\ 0, & t_{on} < t \le T \end{cases} \qquad \text{Eq. 3.25}$$

$$\frac{d}{dt}\begin{bmatrix} i \\ v_c \end{bmatrix} = \mathbf{A}\begin{bmatrix} i \\ v_c \end{bmatrix} + \mathbf{B}\begin{bmatrix} g \\ V_s \\ V_b \\ V_D \end{bmatrix} \qquad \text{Eq. 3.26}$$

$$A = \begin{bmatrix} \dfrac{-\left(R_s + R_L + R_D + \dfrac{R_b R_c}{R_b + R_c}\right)}{L} & \dfrac{-\left(\dfrac{R_b}{R_b + R_c}\right)}{L} \\ \dfrac{\left(\dfrac{R_b}{R_b + R_c}\right)}{C} & \dfrac{-\left(\dfrac{1}{R_b + R_c}\right)}{C} \end{bmatrix}$$

Eq. 3.27

$$B = \begin{bmatrix} B_{11} & \dfrac{1}{L} & \dfrac{-\left(\dfrac{R_c}{R_b + R_c}\right)}{L} & \dfrac{-1}{L} \\ \dfrac{-\left(\dfrac{R_b}{R_b + R_c}\right)i}{C} & 0 & \dfrac{\left(\dfrac{1}{R_b + R_c}\right)}{C} & 0 \end{bmatrix}$$

where

Eq. 3.28

$$B_{11} = \frac{\left(-R_{on} + R_D + \dfrac{R_b R_c}{R_b + R_c}\right)i}{L} + \frac{\left(\dfrac{R_b}{R_b + R_c}\right)v_c}{L}$$
$$+ \frac{\left(\dfrac{R_c}{R_b + R_c}\right)V_b}{L} + \frac{(V_D - V_T)}{L}$$

3.2.2   Canonical Model

The canonical model is the easiest way to obtain the BBC transfer functions using conventional linear circuit analysis methods. The word canonical refers to systems that have a kind of uniqueness in their physical properties. This definition applies to all DC-DC converters including boost converters. First, they transform the voltage and current levels. Second, they contain low-pass filtering of the waveforms. Third, the converter waveforms can be controlled by variation of the duty cycle [38]. The canonical circuit can be configured using the small-signal AC equivalent circuit of the CCM boost converter [1]. This circuit allows to extract physical insight, and to compare the AC properties of converters [60]. The canonical circuit contains frequency-dependent voltage and current sources and effective low pass filter with effective inductance and capacitance and effective resistance losses coupled via ideal DC and AC transformer, as shown in

Figure **3.15**. The complete derivation of the canonical model for a non-ideal BBC is detailed in the appendix. The canonical model parameters for the non-ideal BCC are listed in Table 3.1. These parameters show that the effective values depend on both the physical values and the operating point.

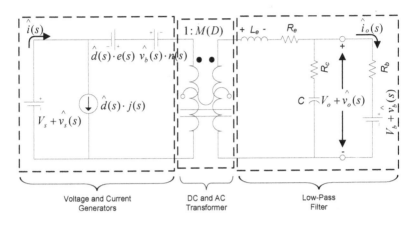

Figure 3.15:   Canonical model for non-ideal BBC

Table 3.1:   Canonical model parameters for non-ideal BBC

Parameter	Value	Parameter	Value
V_{ee}	$-V_T + V_D + \dfrac{(R_c V_b + R_b V_c)}{(R_b + R_c)}$ $+ I\left(-R_{on} + R_D + \dfrac{R_b R_c}{(R_b + R_c)}\right)$	$j(s)$	$\dfrac{I}{(1-D)}$
R_{ee}	$R_s + R_L + D R_{on}$ $+ (1-D)\left(R_D + \dfrac{R_b R_c}{(R_b + R_c)}\right)$	$e(s)$	$V_{ee} - \dfrac{I(sL + R_{ee})}{(1-D)}$
L_e	$\left(M(D)\right)^2 L$	$n(s)$	$\dfrac{R_c(1-D)}{(R_b + R_c)}$
R_e	$\left(M(D)\right)^2 R_{ee}$	$M(D)$	$\dfrac{(R_b + R_c)}{R_b(1-D)}$

Having the canonical model parameters at specific quiescent operating points, the input-to-output transfer function is obtained by setting both the duty cycle variations and the battery voltage variations to zero and the control-to-output transfer function is obtained by setting both the input voltage variations and the battery voltage variations to zero, as in Eq. 3.29 and Eq. 3.30, respectively.

Similarly, the battery-to-output transfer function is obtained by setting the variations in the duty cycle and the input voltage to zero and the output impedance transfer function is obtained by setting the variations in the input voltage, the duty cycle and the battery voltage to zero, as in Eq. 3.31 and Eq. 3.32, respectively. The transfer functions parameters are summarized in Table 3.2. Knowing the exact transfer functions helps defining the effect of varying the inductance value on the converter's behavior and its controllability, as illustrated in sections 4.1 and 6.2 afterward.

$$G_{vs}(s)\Big|_{\substack{\hat{d}(s)=0 \\ \hat{v}_b(s)=0}} = \frac{\hat{v}_o(s)}{\hat{v}_s(s)} = G_{vs_0} \frac{1}{1 + \dfrac{s}{Q\omega_o} + \left(\dfrac{s}{\omega_o}\right)^2} \qquad \text{Eq. 3.29}$$

$$G_{vd}(s)\Big|_{\substack{\hat{v}_s(s)=0 \\ \hat{v}_b(s)=0}} = \frac{\hat{v}_o(s)}{\hat{d}(s)} = G_{vd_0} \frac{-\dfrac{s}{\omega_z} + 1}{1 + \dfrac{s}{Q\omega_o} + \left(\dfrac{s}{\omega_o}\right)^2} \qquad \text{Eq. 3.30}$$

$$G_{vb}(s)\Big|_{\substack{\hat{v}_s(s)=0 \\ \hat{d}(s)=0}} = \frac{\hat{v}_o(s)}{\hat{v}_b(s)} = G_{vb_0} \frac{\dfrac{s}{\omega_b} + 1}{1 + \dfrac{s}{Q\omega_o} + \left(\dfrac{s}{\omega_o}\right)^2} \qquad \text{Eq. 3.31}$$

$$Z_{out}(s)\Big|_{\substack{\hat{v}_s(s)=0 \\ \hat{d}(s)=0 \\ \hat{v}_b(s)=0}} = \frac{\hat{v}_o(s)}{\hat{\imath}_o(s)} = Z_o \frac{\dfrac{s}{\omega_i} + 1}{1 + \dfrac{s}{Q\omega_o} + \left(\dfrac{s}{\omega_o}\right)^2} \qquad \text{Eq. 3.32}$$

Table 3.2:   Transfer functions parameters for non-ideal BBC

Parameter	Value	Parameter	Value
ω_o	$\sqrt{\dfrac{R_b + R_c + R_e}{(R_b + R_c)L_e C}}$	G_{vso}	$\dfrac{M(D)}{(1 + \dfrac{R_e}{(R_b + R_c)})}$
ω_z	$\dfrac{1}{L}\left(\dfrac{(1-D)V_{ee}}{I} - R_{ee} \right)$	G_{vdo}	$\dfrac{1}{(1-D)}(V_{ee} - IR_{ee})$
ω_b	$\dfrac{R_e(R_b - R_c)}{R_b L_e} - \dfrac{R_c(R_b + R_c)}{R_b L_e}$	G_{vbo}	$\dfrac{R_e(R_b - R_c) - R_c(R_b + R_c)}{R_b(R_b + R_c + R_e)}$
ω_i	$\dfrac{R_{ee}}{L}$	Q	$(R_b + R_c)\dfrac{\sqrt{\left(1 + \dfrac{R_e}{(R_b + R_c)}\right)L_e C}}{L_e + (R_b + R_c)R_e C}$

3.3   Maximum Energy Transfer in DC-DC Converters

This section presents the principle of maximum energy transfer (MET) concept based on dynamic inductor control. The MET concept meant to deal with the amount of energy transferred from the input to the output of the BBC and the inductance value effect on the transferred energy. This concept could be implemented either to increase the output power at a certain operating point of the converter or to improve the efficiency by a certain output power.

Recalling the input-to-output transfer function in Eq. 3.29 obtained in section 3.2.2, the quality factor of the BBC converter is defined as in Eq. 3.33. For the sake of clarity, the MET concept is explained for ideal BBC by setting all the conduction losses resistances and voltages to zero. Then, the BBC quality factor connected to the load $R = V_o/I_o$ becomes as depicted in Eq. 3.34.

$$Q = (R_b + R_c) \frac{\sqrt{\left(1 + \frac{R_e}{R_b + R_c}\right) L_e C}}{L_e + (R_b + R_c) R_e C} \qquad \text{Eq. 3.33}$$

$$Q = \frac{1}{2\zeta} = (1 - D) R \frac{\sqrt{C}}{\sqrt{L}} \qquad \text{Eq. 3.34}$$

The quality factor of the system is generally defined as the ratio of stored energy to energy dissipated per cycle [38]. In the BBC, the dissipated energy per cycle is the energy delivered to the load R. Accordingly, smaller quality factor transfers more energy to the output. From Eq. 3.34, if the load and the duty cycle are kept constant, the quality factor can be controlled to transfer more energy to the output by increasing the inductance value L.

Higher inductance value L transfers more energy to the output but at the same time slows the BBC dynamic response against load change. It is well known from the control theory that the critically damped system converges to steady-state position as fast as possible without oscillating. Accordingly, the maximum energy transfer without losing system's dynamics occurs at maximum inductance value L_c that sets the system in critically damping response. Since a critically damped system has a quality factor of 0.5, L_c can be calculated from Eq. 3.34 as in Eq. 3.35.

$$L_c = 4(1 - D)^2 R^2 C \qquad \text{Eq. 3.35}$$

The average output values of the BBC remains constant when enough energy is transferred per second (or cycle) from its input to its output. When the transistor is on during MODE I, the inductor should store an energy that will be fed to the capacitor and load during MODE II and to compensate for the energy removed from capacitor in MODE I.

Assuming a BBC with a negligible ripple value and constant average output voltage and current of V_o and I_o, respectively, the required transfer energy per cycle for keeping the output values constant is shown in Eq. 3.36. The energy amount that the inductor L stored when the transistor switched on and then delivered when it is switched off is depicted in Eq. 3.37, and should equal E_r to keep the output values constant. By substituting Eq. 3.16 into Eq. 3.37, then equating it to Eq. 3.36, the minimum inductance value for transferring the required amount of energy can be calculated as in Eq. 3.38.

$$E_r = \frac{V_o\, I_o}{f}$$

Eq. 3.36

$$E_a = \frac{1}{2}LI^2$$

Eq. 3.37

$$L_{MET} = \left((1-D)\left(\frac{R_b}{R_b + R_c}\right) \right)^2 \frac{2V_o}{f\,I_o}$$

Eq. 3.38

When the inductance value is increased while the duty cycle and the switching frequency are kept constant, more energy will be delivered to the load. It is worth to mention that the input power stays constant and only the amount of energy transferred from the input to the output of the BBC is increased.

The minimum inductance L_{CCM} required for CCM operation for an ideal BBC is shown in Eq. 3.39 [25]. By setting all the conduction losses resistances and voltages in Eq. 3.38 to zero and then substituting Eq. 3.39 into it, the MET inductance for ideal BBC is found to as in Eq. 3.40. The Eq. 3.40 depicts that the minimum inductance value that keeps the BBC in CCM with negligible ripple and RMS losses is at least four times bigger than the CCM inductance L_{CCM}.

$$L_{CCM} = \frac{D(1-D)^2 R}{2f}$$

Eq. 3.39

$$L_{MET} = \frac{4}{D} L_{CCM} \qquad \text{Eq. 3.40}$$

The largest CCM inductance occurs with the lightest load (e.g. R is high) and at duty cycle closest to one-third, as depicted in Eq. 3.41. Substituting Eq. 3.41 into Eq. 3.40 gives the minimum MET inductance value, as in Eq. 3.42. In order to keep operating in the MET where the conduction and RMS losses are minimized and the system response is fast enough for the load changes through the whole operating range $D_{min} \le D \le D_{max}$, the minimum MET inductance value can be defined as in Eq. 3.43. The equation depicts that wider operating ranges require higher inductance value to stay working within MET. Eq. 3.43 depicts that the maximum L_{MET} occurs at the minimum duty cycle value, where the duty cycle is inversely proportional to the input voltage V_s (see Eq. 3.21). The MET concept is assessed in section 5.5 in order to evaluate the boundaries to achieve high efficiency without losing the system's dynamics.

$$\frac{dL_{CCM}}{dD} = \frac{R}{2f}(D-1)(3D-1) = 0$$

$$\xrightarrow{\text{yields}} L_{CCM} = \begin{cases} \dfrac{2R}{27f} & , \quad D = \dfrac{1}{3} \\ 0 & , \quad D = 0 \end{cases} \qquad \text{Eq. 3.41}$$

$$L_{MET} = \frac{8R}{27Df} \qquad \text{Eq. 3.42}$$

$$L_{MET} = \frac{8R}{27D_{min}f} \qquad \text{Eq. 3.43}$$

4 Dynamic Inductor Control Concept

This chapter is conducted to demonstrate the deployment of the dynamic inductor control in the DC-DC converters to achieve the maximum efficiency under different operating conditions. The dynamic inductor control sets the inductor at its maximum inductance value by controlling magnetic field intensity H using a DC control current to keep the inductor working away from its saturation point and forcing it to operate at the maximum possible permeability point μ_{max} (maximum inductance value).

The inductance control improves the converter's efficiency at all operating conditions including variable (heavy and light) load, variable input voltage and variable switching frequency through keeping the maximum possible inductance value and thus reducing RMS and conduction losses. The advantages of the proposed novel dynamic inductor control are summarized at the end of this chapter. It is worth to mention that the maximum values (e.g. maximum inductance or permeability) are the maximum "possible" values under the current operating conditions and not necessarily the absolute maximum values.

4.1 Inductance Value Effect on Efficiency

The efficiency of the boost DC-DC converter is associated with the inductance value [61]. The BBC losses are mainly identified by the RMS losses and the conduction losses, as demonstrated in section 2.4 in chapter 2. The RMS losses and the conduction losses have a nested relationship. The RMS current I_{L_RMS} and the losses P_{loss} are inversely proportional to the inductance value L, as depicted in Eq. 4.1 and Eq. 4.2, respectively.

$$I_{L\_RMS} = \sqrt{I_{av}^2 + \left(\frac{DV_s}{2\sqrt{3}Lf}\right)^2} \qquad \text{Eq. 4.1}$$

© Springer Fachmedien Wiesbaden GmbH, part of Springer Nature 2019
O. Abu Mohareb, *Efficiency Enhanced DC-DC Converter Using Dynamic Inductor Control*, Wissenschaftliche Reihe Fahrzeugtechnik Universität Stuttgart, https://doi.org/10.1007/978-3-658-25147-5_4

$$P_{loss} = \left(I_{av}^2 + \left(\frac{DV_s}{2\sqrt{3}Lf} \right)^2 \right) R_{ee} \qquad \text{Eq. 4.2}$$

Figure 4.1 shows the nested effect between the inductance value and the losses from normalized experimental measurements of BBC. At time t_1, the inductance value L is increased and the duty cycle D is kept constant. The average input current I_{av} remains the same and the RMS current I_{L_RMS} will be reduced and thus the BBC losses will be reduced as well. The reduction in losses may allow more energy to be transferred to the output. In a case of battery load as in Figure 4.1, the output voltage is constant during many switching periods and thus the output current I_o and the efficiency will increase at time t_1.

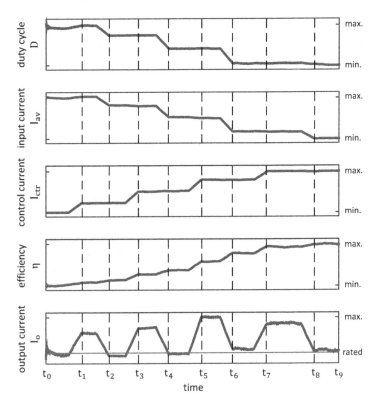

Figure 4.1:   Nested effect of BBC losses

The increment in output current I_o is encountered by reduction in duty cycle D at time t_2 in order to restore the rated output current. The conduction losses are inversely proportional to the duty cycle, as in Eq. 3.17 and Eq. 3.18. Reducing the duty cycle D will reduce the conduction losses and average input current I_{av}. Accordingly, the BBC losses P_{loss} will further decrease and the efficiency will increase at time t_2. This process will be repeated by the dynamic inductor control until the maximum inductance value and maximum efficiency at the rated output current are reached at time t_9.

4.2   Variable Inductor Structure

The dynamic inductor control could be done by adding a secondary winding to the toroid core inductor in Figure 2.1, which is used to force the magnetic field intensity H to be at the maximum permeability point. The variable inductor control is made from one magnetic core and two windings. It consists of a DC control current circuit and a load current circuit, which are connected by a magnetic circuit, as shown in Figure 4.2.

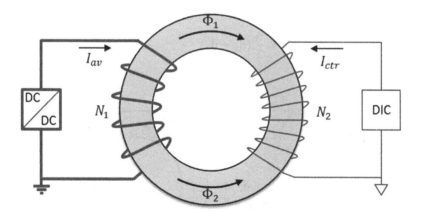

Figure 4.2:   Variable inductor

The DC control current flows through the additional secondary winding N_2 and establishes a magnetomotive force $\mathcal{F}_2$ and a constant DC flux Φ_2. The load current flowing in the primary winding N_1 is actually the input current of the DC-DC converter, which establishes a magnetomotive force $\mathcal{F}_1$. Since this current is a DC current with a superimposed ripple changing in magnitude, it produces a DC flux Φ_1 that is slightly changing in magnitude. A toroid ferrite core is used to construct the variable inductor due to its lower resistivity and higher permeability and saturation flux density, making it a good choice for switched-mode applications below 1 MHz [23, 62, 63]. A non-unity turns ratio of $N_1 : N_2$ is used with higher turns number on the secondary winding to reduce the required control current. The main constraints for the maximum number of windings are the core's geometry and the coil dissipative losses. Using superconducting wires or increasing the winding wire gauge are possible ways to minimize this loss, but only with considerable increases in cost, size, and weight.

The effective mean magnetic path length l_e and effective cross-sectional area A_e for toroid core with uniform rectangular cross-section in Figure 4.3 are found according to Eq. 4.3 and Eq. 4.4, respectively [19].

$$l_e = \frac{\pi (d_1 - d_2)}{\ln\left(\frac{d_1}{d_2}\right)} \qquad \text{Eq. 4.3}$$

$$A_e = \frac{h}{2} (d_1 - d_2) \qquad \text{Eq. 4.4}$$

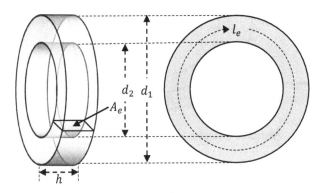

Figure 4.3:   Core structural diagram

Both of primary and secondary electric circuits are connected via a magnetic circuit (the toroid core), so that operating characteristics of any circuit affect the operation of all the interconnected circuits [17]. From this point on, the magnetomotive force and flux produced by the DC-DC converter input current are cited as load magnetomotive force $\mathcal{F}_1$ and flux Φ_1, and the magnetomotive force and flux produced by the DC control current are cited as control magnetomotive force $\mathcal{F}_2$ and flux Φ_2. The variable inductor control operation is described in the following:

■ **Inactive control loop:** the DC control current is set to zero ($I_{ctr} = 0$) and only the DC-DC converter input current I_{av} is flowing through the primary windings N_1 causing a load flux Φ_1 to flow in the core. As a result, the inductor to operate at a certain magnetic field intensity H_1 and permeability μ_1 which is beyond the maximum permeability point μ_{max} of the inductor, as shown in Figure 4.4.

■ **Active control loop:** while the DC-DC converter input current I_{av} is flowing through the primary windings N_1, the DC control current is increased ($|I_{ctr}| > 0$) to flow through the secondary windings N_2 causing a magnetic field intensity H_2 and, accordingly, a control flux Φ_2 that aids or opposes the load flux Φ_1. Assuming that the maximum permeability point μ_{max} of the inductor occurs at the magnetic field intensity H_{max}. If the control current I_{ctr} is set to a value that produces control flux Φ_2 opposes (or aids) the load flux Φ_1, as shown in Figure 4.2, the inductor can be forced to operate at the maximum permeability point μ_{max} and maximum inductance, as in Eq. 4.5. By substituting Eq. 2.3 and Eq. 2.5 into Eq. 4.5, it can be seen that for each DC-DC converter input current I_{av} there is a certain DC control current I_{ctr} which keeps the inductor operating at the maximum permeability point μ_{max}, as in Eq. 4.6.

$$H_{max} = H_1 - H_2 \qquad\qquad \text{Eq. 4.5}$$

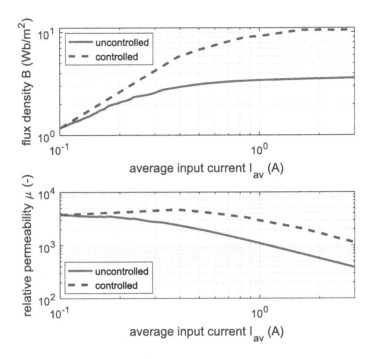

Figure 4.4:   Variable inductor operation principle

$$\frac{B_{max}}{\mu_{max}} = \frac{N_1 I_{av} - N_2 I_{ctr}}{l_e} \qquad \text{Eq. 4.6}$$

The control current is used to drive the magnetic core away from saturation, thereby changing the permeability of the core to its maximum possible value as in Figure 4.4. This changes the inductance to its maximum value at different DC-DC converter input current I_{av} effectively. The measured inductance versus control current at different DC-DC converter input current is shown in Figure 4.5. The dashed line in the Figure 4.5 indicates the controlled maximum inductance value for different DC-DC converter input current as stated in Eq. 4.6. Figure 4.6 confirms the relationship in Eq. 4.6, so that for each average input current there is a certain DC control current which keeps the inductor operating at the maximum inductance.

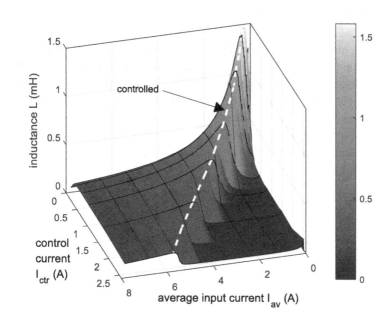

Figure 4.5:   Inductance at different average input current and control current

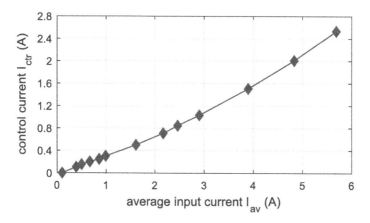

Figure 4.6:   Control current vs. average input current

4.3   Control Methodology

The variable inductor control with DC-DC converter is shown in Figure 4.7 and it is in the following described. Two control modes (constant current and constant voltage) can be distinguished in the DC-DC converter with a battery on the output as a load. In the constant current (CC) mode, the output current I_o remains constant while the output voltage V_o is slowly building up. On contrary, the output voltage V_o remains constant while the output current I_o is slowly dropping out during the constant voltage (CV) mode.

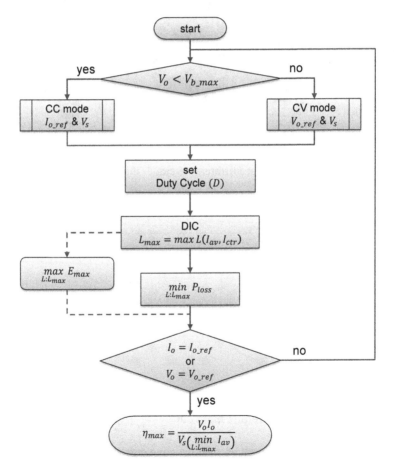

Figure 4.7:   Variable inductor control with DC-DC converter control

The change in the output current and voltage values is very slow compared to the switching frequency of DC-DC converter and can be constant considered during several switching cycles. This means that the DC-DC converter has a predefined operating point at constant input supply voltage V_s, output voltage V_o and output current I_o.

The inductor has a certain inductance value of $L(I_{av})$, as depicted from Eq. 2.7, at this operating point. If a control current I_{ctr} is set to a certain value, the inductor can be forced to operate at maximum inductance, so $L(I_{av}, I_{ctr}) = L_{max}$. Forcing the inductor to operate at its maximum value has two effects: reducing the conduction losses P_{loss} or alternatively increasing the energy transferred from the DC-DC converter input to its output E_{max}, as explained earlier in section 4.1.

The reduced conduction losses or the increased energy transfer lead to higher output voltage or current value at the currently-set duty cycle. In order to restore the original output voltage and current, the DC-DC converter controller will reduce the duty cycle D and, thus, input average current I_{av}, as depicted in Eq. 3.16. The reduction in D and I_{av} is offset by higher efficiency at the same original input supply voltage V_s, output voltage V_o and output current I_o values (refer to Eq. 3.24).

The control strategy sets the inductance value at initial inductance L_o by setting the initial control current to be greater than I_{ctr_max}, which is defined according to Eq. 4.7. The value H_{max} can be found from the inductor's datasheet or from experimental measurement.

The initial control current I_{ctro} can be defined as a function of the average input current I_{av} as in Eq. 4.8. The limits of factor B_{lim} can be found as follow: assume that the maximum inductance value occurs at $H_{max}l_e = 0$. By substituting in Eq. 4.7, the factor B_{lim} is equal to 1, as depicted in Eq. 4.9, where the highest control current is required. Now assume that the maximum inductance value occurs at $H_{max}l_e = A$, where $0 < A < N_1 I_{av}$. By substituting in Eq. 4.7, the factor B_{lim} is less than 1, as depicted in Eq. 4.10.

$$I_{ctr\_max} = \frac{N_1 I_{av} - H_{max}l_e}{N_2} \qquad \text{Eq. 4.7}$$

$$I_{ctro} = B_{lim} \frac{N_1}{N_2} I_{av} = \frac{I_{av}}{N} \qquad\qquad \text{Eq. 4.8}$$

$$B_{lim} \frac{N_1}{N_2} I_{av} = \frac{N_1 I_{av} - 0}{N_2} \quad\xrightarrow{\text{yields}}\quad B_{lim} = 1 \qquad\qquad \text{Eq. 4.9}$$

$$B_{lim} \frac{N_1}{N_2} I_{av} = \frac{N_1 I_{av} - A}{N_2} \quad\xrightarrow{\text{yields}}\quad B_{lim} = \left(1 - \frac{A}{N_1 I_{av}}\right) < 1 \qquad \text{Eq. 4.10}$$

Accordingly, the factor B_{lim} limits are found to be within $(1 - A/(N_1 I_{av}))$ and 1. It is worth mentioning that an aiding control current is required when $A > N_1 \, I_{av}$. It is recommended to set factor B_{lim} to 1 when the maximum inductance point A is not exactly known in order to cover the whole possible inductance range.

The dynamic inductor control DIC algorithm in Figure 4.7 is detailed in Figure 4.8 as follow:

- **Set initial control current:** at the first iteration ($k = 1$), the control current of I_{ctr} and maximum allowed control current I_{max} are set to be I_{avo}/N. The value I_{avo} is the inductor current value before starting the DIC. The applied control current leads to lower the average input current I_{av}.

- **Find minimum inductor current:** after the first iteration, the difference between the average input current and the maximum allowed control current ΔI_L is always checked to be less than a small positive factor $+x$. This positive factor prevents extreme oscillations in the control current. The control algorithm keeps lowering the control current until ΔI_L is slightly greater than $+x$, where the maximum inductance value is reached. Then, the control algorithm keeps oscillating around that point.

- **Limit control current:** the control current is always checked not to exceed the average inductor current at any time.

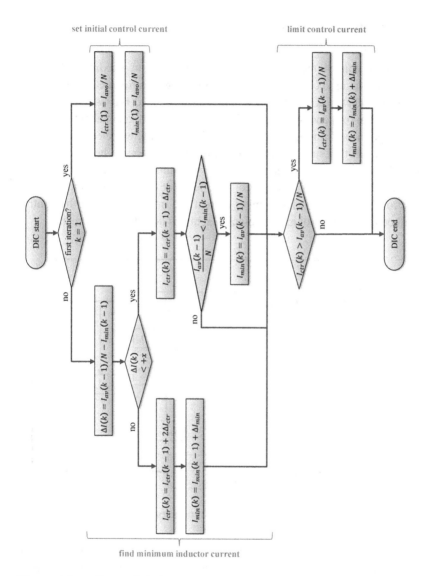

Figure 4.8:   Dynamic inductor control algorithm

4.4   Stability Investigation

Lyapunov control method has been implemented to maximize the BBC efficiency at any certain operating point. The control method treats the coupled inductor as a stand-alone automatically tuned filter similar to those described in [64] and [65]. This method extracts information about a function to be optimized from perturbation signal injected in power electronic system, so that a function such as input current or input power can be minimized by correlating the perturbation signal itself with this function, and then using this correlation to adjust the control parameter [66]. For the dynamic inductor control, the average inductor/input current I_{av} is considered for the control method and the control method block diagram is depicted in Figure 4.9.

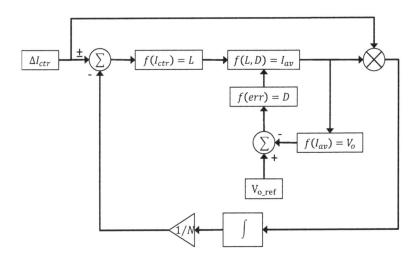

Figure 4.9:   Block diagram of dynamic inductor control

The BBC controller (duty cycle controller) is set to have much faster response than dynamic inductor controller so that the duty cycle can be in quasi-steady state. Then, the control method block diagram can be simplified as in Figure 4.10. The controller measures the average inductor/input current I_{av} of the BBC and tunes the inductance value L to minimize the ripple factor r and average input current I_{av}, and thus maximize the efficiency. The average input current I_{av} value is obtained using a low-pass filter that filters all the noises with a frequency higher than the frequency of the perturbation signal ΔI_{ctr}.

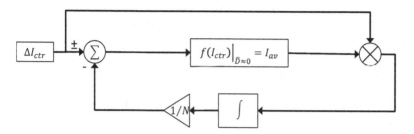

Figure 4.10:   Simplified block diagram of dynamic inductor control

The dynamic inductor controller generates a small perturbation signal ΔI_{ctr} in the control current that controls the inductance value L and thus the average input current I_{av}, as shown in Figure 4.11.

It is worth to mention that injecting a small perturbation signal to find the maximum efficiency point is an art of extremum seeking control. This control method is a non-model based real-time optimization approach for dynamic problems where only limited knowledge of a system is available, such as when the system has a nonlinear equilibrium map which has a local minimum or maximum [67].

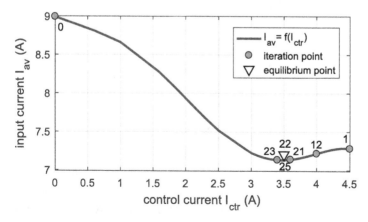

Figure 4.11:   Average input current as a function of control current

Assume the function $f(I_{ctr}) = I_{av}$ of Figure 4.11 as a Lyapunov function of the system over a region $\mathbb{R}$ of the state-space and $\bar{I}_{ctr}$ is an equilibrium point of the system in that region. The control method stability can be proofed using Lyapunov direct method described in [68], when the following three requirements for a Lyapunov function $V(I_{ctr})$ are satisfied: V is continuous, $V(I_{ctr})$ has a unique minimum at $\bar{I}_{ctr}$ with respect to the other points in region $\mathbb{R}$ and the value of V never increases along any trajectory of the system contained in region $\mathbb{R}$.

The experimental measurement of the function $f(I_{ctr}) = I_{av}$ shown in Figure 4.11 confirms the continuity and unimodal behavior of the function on the region of interest. The control strategy in Figure 4.8 is implemented so that the function $f(I_{ctr})$ never increases and tends towards minimum of the associated state variable I_{av}. Accordingly, all the requirements are fulfilled and the control method is stable at the equilibrium point $\bar{I}_{ctr}$.

Table 4.1: Average input and control current at different iterations

Iteration	I_{ctr} (A)	I_{av} (A)	
0	0	9	
1	4.50	7.31	
2	4.45	7.30	
⋮	⋮	⋮	
12	4.00	7.21	
⋮	⋮	⋮	
21	3.55	7.13	
22	3.50	7.13	steady-state
23	3.45	7.15	
24	3.55	7.13	
25	3.50	7.13	

A numerical example with the parameters $I_{avo} = 9\,A$, $B = 1.0$, $\Delta I_{ctr} = 0.05\,A$, $\Delta I_{min} = 3\,\Delta I_{ctr} = 0.15\,A$, and $N_2/N_1 = 2$ is given in Figure 4.11 and Table 4.1 to clarify the control strategy. At the first iteration, a maximum control current $I_{ctr}(1) = I_{avo}/2 = 4.5\,A$ is applied and it leads to a lower average input current of $I_{av}(1) = 7.31$ A. The control strategy keeps lowering the control current by $\Delta I_{ctr} = 0.05$ A until the average input current starts increasing again at 23rd iteration. Then, control algorithm starts slightly increasing and decreasing the control current around $I_{ctr}(k \geq 24) \cong 3.5$ A, so that the function $f(I_{ctr})$ tends always toward its minimum.

4.5 Advantages of Dynamic Inductor Control

The novel dynamic inductor control introduced in this dissertation tends to have many advantages over the prior arts of inductor control in DC-DC converters. The main advantages of the proposed dynamic inductor control are listed below:

- **Generally applicable to conventional DC-DC converters:** the output variables (voltage or current) of DC-DC converter are regulated by their original control means and the proposed dynamic inductor control is only used to control the inductance value. This makes the DIC applicable to many conventional DC-DC converters.

- **Physical inductor parameter adaptation:** the proposed DIC method controls the inductance value at all operating points. Thus, the DC-DC converter can efficiently operate again at its physically optimized conditions without adding or removing any physical component in the system. The inductor parameter adaptation consists of maximizing the efficiency function which is used as control parameter in the DIC.

- **Improved efficiency at different operating points:** the novel DIC does not only archive higher efficiency at both light and heavy loads, but it also guarantees higher efficiency of the converters with varying source voltage, such as battery-powered systems and renewable energy applications.

The previously mentioned advantages are simulative and experimentally proved in the following chapters.

5 Dynamic Inductor Control Simulation

In this chapter, the dynamic inductor control strategy is numerically evaluated to validate its efficiency and stability. MATLAB Simulink model was implemented for the numerical evaluation in section 5.1. The MET concept from section 3.3 is also evaluated in order to illustrate the possible tradeoffs between the systems dynamics and efficiency.

5.1 Controller Simulation

The dynamic inductor control strategy with the non-ideal BBC model was numerically evaluated through MATLAB Simulink with SimPowerSystems toolbox. The simulation model is shown in Figure 5.1 and it has three main parts. These parts are briefly described in the following, where the complete Simulink model with the control strategy is illustrated in the appendix.

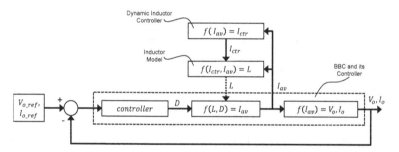

Figure 5.1:   Simulink model of the BBC with dynamic inductor control

■ **BBC model and its controller:** the BBC model in Figure 5.1 was implemented based on the circuitry model developed earlier in chapter 3. The main losses in the BBC are modeled in form of resistors and voltage drops in order to simulate the real converter's dynamics and characteristics. A proportional-integral (PI) controller is used to regulate output voltage or current through sequential switching of the DC power source at certain switching frequency and variable duty cycle D.

© Springer Fachmedien Wiesbaden GmbH, part of Springer Nature 2019
O. Abu Mohareb, *Efficiency Enhanced DC-DC Converter Using Dynamic Inductor Control*, Wissenschaftliche Reihe Fahrzeugtechnik Universität Stuttgart, https://doi.org/10.1007/978-3-658-25147-5_5

▣ **Inductor model:** this model is divided into two sub-models: one is used
to simulate the effects of varying the average inductor/input current I_{av}
and the control current I_{ctr} on the value of the variable inductance L. It
is empirically determined, as shown in Figure 4.5, and represented by
the function $f(I_{ctr}, I_{av})$. This function is modeled with a 2D lockup
table in the Simulink model. The second sub-model represents the
physical inductor circuitry in the BBC model and it is represented by the
$f(L, D)$ function. The inductor in reality is a part of the DC-DC
converter. Thus, the second sub-model is considered a part of the BBC
model in Figure 5.1 as it represents the physical inductor circuitry. The
variable inductor circuitry is modeled as inductance-based controlled
current source according to Faraday's law of induction in Eq. 5.1, where
the inductance value is fed from the 2D lockup table in the first
sub-model.

$$i(t) = \frac{1}{L(I_{ctr}, I_{av})} \int v_L(t) \, d(t) \qquad\qquad \text{Eq. 5.1}$$

▣ **Dynamic inductor controller:** the dynamic inductor control algorithm
described in section 4.3 is represented by the function $f(I_{av})$ in the
Simulink model. The controller uses the average inductor/input current
I_{av} to update the control current value in a way that the dynamic inductor
controller starts controlling the inductance value when the BBC has
reached its steady-state in order not to interfere with the BBC controller.
Therefore, the dynamic inductor control algorithm is called through the
simulation every sampling time to check the control current not to
exceed the average inductor current at any time, but the control current
value is updated every few sampling times once.

The inductor and DIC Simulink models are shown together in Figure 5.2.
The average input current is fed to DIC and the control current is calculated
according to the algorithm described in section 4.3. The control current is
then subtracted from the average control current to find the resultant
magnetic field intensity and the new inductance value from the lockup table
in the first inductor sub-model. The variable inductance value is transformed
throughout Faraday's law of induction to a variable inductor current in the
second inductor sub-model.

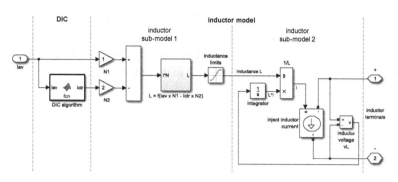

Figure 5.2:   Inductor and DIC Simulink models

5.2   Simulation Assumptions and Parameters

In order to create a reliable simulation model of the non-ideal BBC model, it is usually necessary to make a number of assumptions and approximations about the model being simulated. The interpretation of the simulation results requires a clear understanding of the *assumptions* made in the simulated model. The simulation model is implemented based on the following assumptions: (1) the electric and electronic components parameters are considered time-invariant and constant under all operating conditions and (2) the temperature effect on the electric and electronic components parameters is not considered and is negligible.

The BBC is a stiff ordinary differential equation (ODE) system which the solution for its model consists of rapidly changing parts together with slowly changing parts [69]. Thus, ODE23s solver method, which computes the model's state at the next time step using a modified Rosenbrock formula of order two, is used with the simulation for its adequate accuracy and time efficient.

A lithium-ion polymer battery cell is used as a load for the BBC with nominal capacity of 5.4 Ah, nominal voltage of 3.7 V, voltage range between 2.8 V to 4.2 V and maximum charging rate of 3C [70]. The maximum internal resistance specified in the datasheet is 4 mΩ. The real battery internal resistance was measured as in [71] and the results are shown in Figure 5.3.

As such in [71], [72] and [73], the figure depicts a constant internal resistance value during the charging process and a minor difference of 0.74 mΩ at different charging rates. A constant average value of 10 mΩ is used for the battery internal resistance R_b in the simulation.

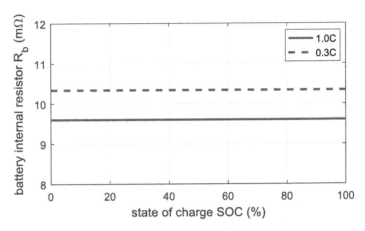

Figure 5.3:   Battery internal resistor at different SOC and charging rates

Table 5.1:   Components simulation parameters

Component	DC power source	Inductor	Power Switches		Capacitor	Battery
Ideal values						
Symbol	Vs	L	MOSFET	Diode	C	Vb
Unit	(V)	(mH)	-	-	(mF)	(V)
Value	2.7 – 3.3	Variable	-	-	6.6	2.8 – 4.2
Internal resistance losses in (mΩ)						
Symbol	Rs	RL	Ron	RD	Rc	Rb
Value	0.001	41	38	260	2.1	10
Voltage drop losses in (V)						
Symbol	-	-	VT	VD	-	-
Value	-	-	0.0	0.15	-	-

The simulation parameters for the electric and electronic components are extracted from the components' datasheets intended for the experimental setup. The specifications of Li-ion battery, resistive load, DC power source, switching frequency, MOSFET and diode are specified first, then used to determine the proper inductor for the BBC according to Eq. 3.43. The parameters for the Li-ion battery, resistive load, DC power source, switching frequency, MOSFET, diode, inductor and capacitor used in the simulation are summarized in Table 5.1. The switching frequency is 20 kHz.

5.3   Simulation Results

The simulation results obtained in this section are to evaluate the effects of the DIC on the BBC steady-state and dynamic behavior. The steady-state and dynamic behavior evaluations are performed at the same operating conditions used in the experimental evaluations in section 7.1 and section 7.2, respectively. This allows the ability to compare between the simulation model prediction and experimental observation, as well as, to evaluate the effectiveness of the DIC.

The steady-state performance is assessed at constant source voltage V_s and output voltage V_o of 3.0 V and 4.2 V, respectively, at different operating points (output currents). The simulated steady-state behavior of the RMS input current, the control current, the inductance value, the input current ripple factor and the converter's efficiency at different operating points are depicted in Figure 5.4 to Figure 5.7.

The simulation results indicate the effectiveness of the DIC and its ability to improve the converter's efficiency at all operating points, throughout keeping the maximum possible inductance value and reducing the ripple and the RMS input current. The simulated steady-state control current shown in Figure 5.8 copes well with the experimental measurements in Figure 4.6.

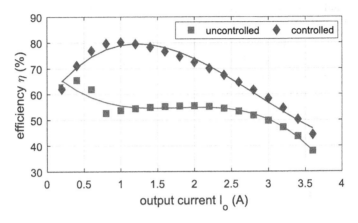

Figure 5.4:   Simulated steady-state efficiency at different output (load) currents

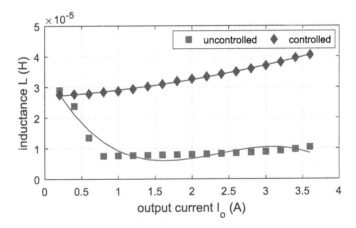

Figure 5.5:   Simulated steady-state inductance values at different output (load) currents

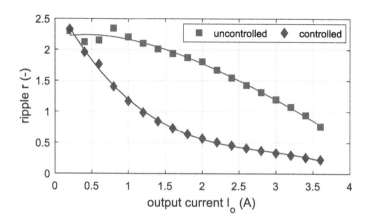

Figure 5.6: Simulated steady-state ripple values at different output (load) currents

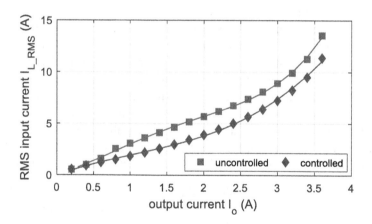

Figure 5.7: Simulated steady-state RMS input current at different output (load) currents

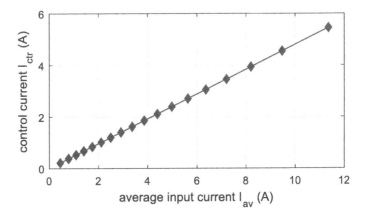

Figure 5.8:   Simulated steady-state control current vs. average input current

In the evaluation of the dynamic behavior performance when the DIC is enabled, the source voltage V_s, the output voltage V_o and the output current I_o are kept constant at 3.0 V, 4.2 V and 2.7 A, respectively. The DC-DC converter is allowed to first reach steady-state, then the DIC is enabled as shown in Figure 5.9. The converter reaches its steady-state again after approximately 52 ms (shaded area), where the maximum inductance value and maximum efficiency for this operating point are achieved. The average input current I_{av} reaches its steady-state value and remains stable at the higher possible inductance value, leading to more than 10 % higher efficiency. The DC-DC converter parameters before and after enabling the DIC is listed in Table 5.2.

The simulation model successfully predicts the effects of implementing the DIC to the DC-DC converter at all operating points. Nevertheless, discrepancies between the simulated and experimental results are noticed. The discrepancies mainly referred to the limitations in the model, due to the lumped models that approximate and simplifies the converter's components behavior under the assumptions described before. The constant values of the resistors and voltage drops at all operating conditions and the unmodeled parasitic effects in the components' models are examples of the models approximation and simplification which result to the discrepancies between the simulated and experimental results shown in chapter 7 afterward.

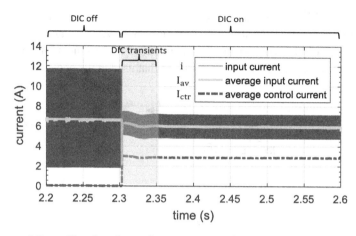

Figure 5.9: Simulated transient response of BBC input current and DIC control current

Table 5.2: Simulation transient parameters with DIC disabled and enabled

Parameter and unit	Symbol	DIC disabled	DIC enabled
Source voltage (V)	V_s	3.00	3.00
Output voltage (V)	V_o	4.20	4.20
Average input current (A)	I_{av}	7.18	5.98
Output current (A)	I_o	2.70	2.70
Inductance (µH)	L	8.62	35.1
Efficiency without DIC losses (%)	η	52.64	63.17

5.4 Evaluating DIC with other DC-DC Converter Type

The buck converter with DIC is evaluated in this section in order to proof the generality of the proposed strategy with the other DC-DC converter types. The buck converter has similar components as in the boost converter, but rather connected together in a different way. The buck converter scheme shown in Figure 5.10 is used in the Simulink model to evaluate its behavior when the DIC is implemented. The source voltage V_s, the output voltage V_o and the output current I_o are kept constant at 5.88 V, 4.2 V and 2.7 A, respectively. The other components parameters used in the simulation are summarized in Table 5.1, and they are similar to that used in the boost converter.

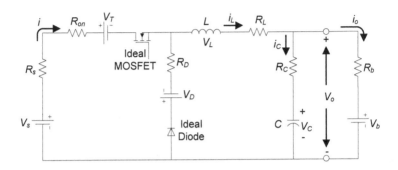

Figure 5.10:   Non-ideal scheme of a buck battery charger

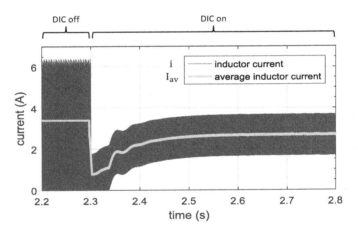

Figure 5.11:   Transient response of buck converter input current as DIC is enabled

The buck converter is allowed to reach steady-state at first, and then the DIC is enabled as shown in Figure 5.11. The converter reaches its steady-state again in about 270 ms, where the maximum inductance is achieved. The conduction losses in the buck converter, as in the boost converter, are inversely proportional to the RMS inductor current [24, 74]. Achieving the maximum inductance value reduces the RMS inductor current and thus improves the converter's efficiency, as seen in Table 5.3.

Table 5.3:   Simulation transient parameters of buck converter

Parameter and unit	Symbol	DIC disabled	DIC enabled
Source voltage (V)	V_s	5.88	5.88
Output voltage (V)	V_o	4.20	4.20
Average input current (A)	I_{av}	3.081	2.444
RMS inductor current (A)	I_L	3.406	2.759
Output current (A)	I_o	2.70	2.70
Inductance (μH)	L	7.00	28.45
Efficiency without DIC losses (%)	η	62.59	78.91

5.5   Maximum Energy Transfer Evaluation

High efficiency performance requires transfer energy per switching cycle to keep the output values constant and with smaller ripple values. But the optimized operating conditions balance between the DC-DC converter circuit performance, cost and size. Over engineering the inductor size results in slow-responsive system behavior, as well as, bulky and expensive design without gaining more efficiency.

The MET concept from section 3.3 sets the boundaries to achieve high efficiency without losing the system's dynamics. In the evaluation of the MET, the source voltage V_s, the output voltage V_o and the output current I_o are kept constant at 3.0 V, 4.2 V and 2.7 A, respectively, and the load $R = V_o/I_o$ equas 1.556 Ω. The duty cycle at these conditions is found using Eq. 3.21 to be around 55.3 %. The MET inductance value L_{MET} is calculated through substituting the previous values into Eq. 3.43 and it is found to be about 41.7 μH.

Increasing the inductance value beyond L_{MET} will dramatically overdamp the converter's response without a remarkable improvement in the efficiency, as depicted in Figure 5.12. Increasing the inductance value by more than three times of L_{MET}, as shown in Figure 5.12, has only improved the efficiency by 0.39 % and caused the converter's settling time to be increased by more than 344 %. The considered settling time here is the time required for the response curve to reach and stay within 2 % of its final value [75].

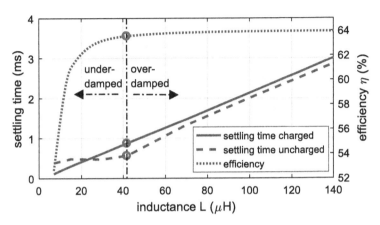

Figure 5.12:   BBC's settling time at different inductance values

The output capacitor C of the BBC is directly connected to the Li-ion battery (load), as shown in Figure 3.9. Accordingly, the capacitor is pre-charged and has an initial voltage equal to the battery voltage V_b. The simulated under-sampled transient behavior of the boost DC-DC converter at different inductance values is shown in Figure 5.13, when the output capacitor is charged and uncharged. The charged capacitor tends to damp the converter's response, even when an underdamped behavior is expected at inductance values less than L_{MET} and the underdamped behavior of the converter is barely noticeable. The damped behavior of the converter in the simulation results matches the experimental results of section 7.2. Beside the overshoot value, the damping influence of a charged output capacitor also affects the settling time of the system.

The step response characteristics of the BBC when the output capacitor is charged and uncharged are listed in Table 5.4. A critically damped and overdamped DC-DC converter with a charged output capacitor has a slower response and a longer settling time compared to that with uncharged output capacitor. The situation is alternated with an underdamped DC-DC converter, where the system needs more time to reach steady-state with uncharged capacitor. Despite the damped behavior of the DC-DC converter with a charged capacitor, only the inductance value around L_{MET} provides high efficiency and dynamic response at the same time.

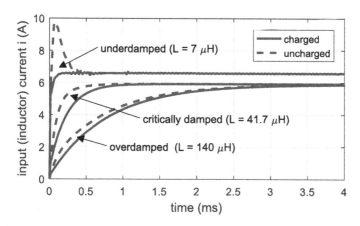

Figure 5.13: Transient behavior of the BBC at different inductance values

The normalized curve at input voltage 3.0 V in Figure 5.14 shows that having an inductance value greater than 70 % of L_{MET} guarantees the efficiency to be more than 99.3 % of that at L_{MET}. This characteristic adds more robustness to the system and enables tolerating the generated perturbation signals of the proposed control strategy in section 4.4.

Table 5.4: Step response characteristics of boost DC-DC converter

Parameter and unit	Underdamped		Critically damped		Overdamped	
	charged	un-charged	charged	un-charged	charged	un-charged
Settling time (ms)	0.111	0.384	0.872	0.563	3.006	2.858
Overshoot (%)	0.904	36.13	0	0	0	0

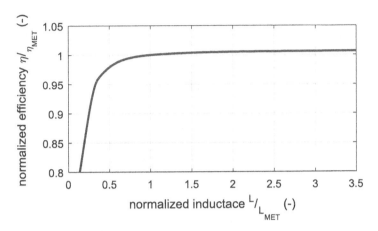

Figure 5.14: Normalized efficiency with respect to MET values at nominal
voltage of 3.0 V

The MET inductance value L_{MET} varies with different loads or input voltage
values. Eq. 3.43 depicts that L_{MET} is proportional to the input voltage V_s,
where the duty cycle value is inversely proportional to the input voltage (see
Eq. 3.21). The proposed DC-DC converter has an input voltage range of
$2.7\,V \le V_s \le 3.3\,V$. The largest L_{MET} value is 49.2 µH and it occurs at an
input voltage of 3.3 V and duty cycle of 0.468. Being the ratio between the
nominal MET inductance at 3.0 V and its counterpart at 3.3 V more than
84.7 %, the converter can still operate at nominal MET inductance
$L_{MET} = 41.7$ µH with an efficiency of around 99.8 % of that at
$L_{MET} = 49.2$ µH.

6 Dynamic Inductor Control Implementation

The dynamic inductor control strategy is implemented with a prototype DC-DC converter to evaluate and validate its effectiveness and stability. The prototype layout is described in section 6.1. The controller implementation on a printed circuit board is detailed in section 6.2.

6.1 Prototype Converter with Dynamic Inductor Control

The proposed DIC strategy of chapter 4 is implemented together with the DC-DC converter on a two-layer printed circuit board (PCB), as shown in Figure 6.1, using standard circuit components listed in Table A.1. The prototype PCB is a standard Eurocard format size of 100 mm × 160 mm and the thickness 1.6 mm. Figure 6.2 shows that the PCB is divided into 3 zones: DC-DC converter is located in zone 1, DIC is located in zone 2 and controller unit for both of them is located in zone 3. The inductor is not fitted on the PCB and can be flexibly replaced and connected to the PCB. Each zone is galvanic isolated from the other zones, but it is possible to connect their grounds to the same potential like in the case of connecting the DC-DC converter and the DIC circuits to the same voltage source. The galvanic isolation is mainly required to safely operate the controller and reduce the noise caused by power circuits, where each circuit is supplied by its own power supply. The isolation is implemented through using isolated gate drivers, magneto-resistive current sensors, opto-isolators and isolated DC-DC converters. The controller area network (CAN) bus is also galvanic isolated and it is located on an external PCB that can be connected via a dedicated socket placed on the upper left corner of the main PCB shown in Figure 6.2.

© Springer Fachmedien Wiesbaden GmbH, part of Springer Nature 2019
O. Abu Mohareb, *Efficiency Enhanced DC-DC Converter Using Dynamic Inductor Control*, Wissenschaftliche Reihe Fahrzeugtechnik Universität Stuttgart, https://doi.org/10.1007/978-3-658-25147-5_6

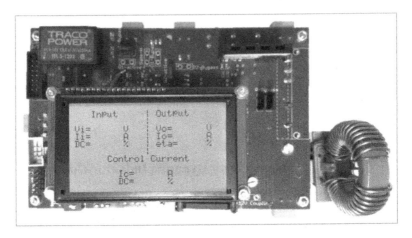

Figure 6.1:   Prototype printed circuit board

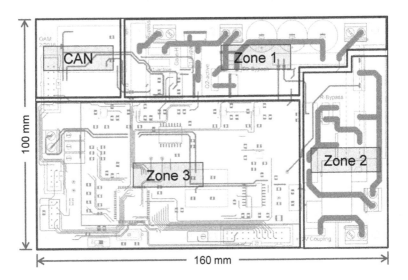

Figure 6.2:   BBC with DIC PCB layout

The DC-DC converter in zone 1 is configurable and has two modes as depicted in Figure 6.3. It can be configured as boost converter (solid-line) by connecting the black-colored jumpers and opening all the gray-colored jumpers or as a buck converter (dashed-line) by connecting the gray-colored jumpers and opening all the black-colored jumpers. The input and output voltage and current values are measured with the same sensors (A1 and A2 in Figure 6.3 are current sensors) and one isolated gate driver is used in both modes. Isolated magneto-resistive current sensors are used to prevent the generated losses when shunt resistors are used to measure currents.

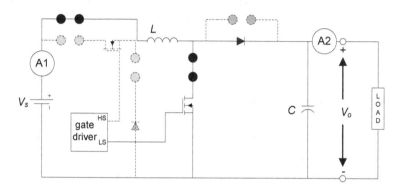

Figure 6.3: Configurable DC-DC converter layout (as boost converter)

The DIC circuitry in zone 2 is shown in Figure 6.4. It uses the PWM technique to set the control current in order to keep a simple design of the DIC and reduce the DIC losses P_{ctr}. The control current is also measured using an isolated magneto-resistive current sensor (A3).

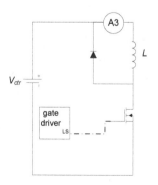

Figure 6.4: DIC layout

The controller unit consists mainly of a microcontroller, isolated analog to digital converter (ADC), isolated DC-DC supplies, graphic liquid-crystal display (LCD) and CAN bus board. The function of the CAN bus board is to transfer the microcontroller internal values via SPI to a high-speed CAN transceiver. Accordingly, the values can be noticed on the graphic LCD and logged on the computer at the same time. The detailed circuits' schematics for the prototype converter with dynamic inductor control are included in the appendix.

6.2   Controller Implementation

The overall system of the BBC augmented with the novel dynamic inductor control is realized using additional circuitry and current sensors. The additional current sensors are required to know the actual operating point. The information provided by current sensors are then used by the additional circuitry to control the inductance of the modified variable inductor and improve the efficiency at all operating points.

Figure 6.5 illustrates the block diagram of a BBC with variable inductor control in accordance with the dynamic inductor control strategy. The main layout and components are in the following described. The BBC includes a switch Q1 for selectively applying input supply voltage V_s to the primary winding of the inductor N_1 to store energy in it. The switch Q1 is a N-MOS transistor which receives a driving signal PWM1 from a N-MOS driver. The diode D1 functions as a switch which is only conducting when switch Q1 is open. It allows a portion of the previously stored energy in the inductor L to be absorbed by the load and the capacitor C1 together with the energy coming from the input supply voltage V_s. Capacitor C1 stores energy supplied from the inductor L in a conventional manner and drives the load in the absence of a driving current I_{av} from the inductor L.

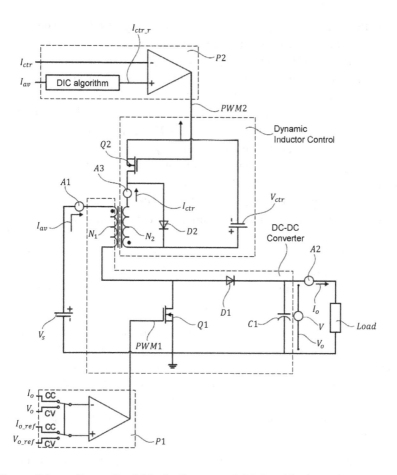

Figure 6.5: Generalized block diagram of BBC with dynamic inductor control

The dynamic inductor control block diagram in Figure 6.5 includes a switch Q2 for controlling the control current I_{ctr} passing through secondary winding of the inductor N_2. Switch Q2 is a N-MOS transistor which receives a driving signal PWM2 from a N-MOS driver which is generated in accordance with the strategy of the dynamic inductor control for reducing the input average current I_{av} and maximizing the inductance value L at the primary side. The diode D2 functions as a rectifying circuit which freewheels the inductor control current I_{ctr} when switch Q2 is open.

Two PI controllers are implemented on a microcontroller to control the output of the BBC and the inductance value. In the BBC controller P1, a control signal PWM1 representing the duty cycle D for regulating the BBC output current/voltage. The control signal PWM1 is generated by comparing the output current I_o/output voltage V_o detected across current sensor A2/voltage sensor V with a reference value I_{o_ref}/V_{o_ref}. In the dynamic inductor controller P2, a control signal PWM2 representing the duty cycle for regulating the inductor control current I_{ctr}. The control signal PWM2 is generated by comparing the inductor control current I_{ctr} detected across current sensor A3 with the control current reference value I_{ctr_r}. The reference value I_{ctr_r} is generated in accordance with DIC algorithm in section 4.3, which minimizes the input average current I_{av} detected across current sensor A1 and maximizes the inductance value. The two controllers cannot conflict at any point where the DC-DC converter controller P1 is made with higher priority than the dynamic inductor control P2. In addition, the inductor control current I_{ctr} is only set when the DC-DC converter controller P1 is in steady-state, as described in section 4.4.

A digital microcontroller is used to implement the previously described control strategy and to regulate the BBC output values. The dynamic inductor control causes low-frequency perturbations of the BBC output value. A proportional-integral (PI) controller is used to regulate the BBC output, where the low-frequency loop gain is increased and the BBC output is better regulated at DC and at frequencies well below the loop crossover frequency [38].

Figure 6.6 shows the BBC control block diagram used to design the digital PI controller and then assess the performance and the stability of the converter. The digital signal blocks are the blocks within the dashed-line in Figure 6.6. The well-regulated output values are always equal to the set-values. The set-values are used as a reference for the converter. The reference values in the CC and CV are I_{o_ref} and V_{o_ref}, respectively.

The delay caused by sampling the feedback signals and calculating the control effort is an additional phenomenon that must be taken into account while implementing a digitally-controlled power supply [76]. There are four specific blocks that enable the digital controller to achieve the high-performance regulation requirements of the BBC: the analog to digital converter (ADC) used to sample the sensor signal, the sensor signal filter at the input of the ADC, the execution time of the microprocessor and the digital pulse width modulator (DPWM) that converts the sampled, compensated error signal into the gate-drive signals.

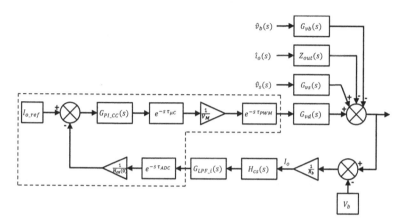

Figure 6.6:   Block diagram of the BBC control model in CC mode

The sensor transfer function is digitally scaled to always have a unity DC gain transfer function with a delay caused by the ADC and its low-pass filter (LPF). The current sensor itself has an additional delay caused by its circuitry. It does not change output state immediately when an input parameter change occurs, as it will change to the new state over a period of time called the response time T_r. The response time is the time required for a sensor output to change from its previous state to a final settled value [77]. It is defined as the time it takes the sensor to reach 90 % of its steady-state value after the introduction of the measurand [78]. The current sensor response time is given in its datasheet [79] and shown in Figure 6.7.

The delay functions can be represented as in Eq. 6.1, where τ is the delay time in seconds [80]. The unity current sensor and the low-pass filter transfer functions are given in Eq. 6.2 and Eq. 6.3. The presence of the input filter and the ADC dramatically increases the sensor's response time as shown in Figure 6.7. The delay times related to implementing digital controller are summarized in Table 6.1. The corresponding MATLAB code used to analyze the transfer functions of Figure 6.7 is presented in the appendix.

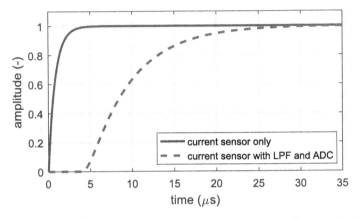

Figure 6.7:   Step response of current sensor, low-pass filter and analog to digital converter

$$\mathcal{L}\{f(t-\tau)\} = e^{-s\tau}\, F(s) \qquad\qquad \text{Eq. 6.1}$$

$$\frac{H_{cs}(s)}{H_{cs}(0)} = \frac{1}{8.4 \times 10^{-7}s + 1} \qquad\qquad \text{Eq. 6.2}$$

$$G_{LPF\_i}(s) = \frac{1}{5.1 \times 10^{-6}s + 1} \qquad\qquad \text{Eq. 6.3}$$

Table 6.1:   Delay times related to digital controller

Device	Symbol	Value in (s)
Analog to digital converter	τ_{ADC}	4×10^{-6}
Microcontroller	$\tau_{\mu C}$	6.25×10^{-7}
PWM	τ_{PWM}	16.5×10^{-6}

7 Experimental Results

This chapter presents the experimental results for the dynamic inductor control implemented in the prototype DC-DC converter to experimentally validate the proposed control strategy. The steady-state and the transient performance of the converter with DIC are demonstrated throughout different load and input transient tests in the following sections.

7.1 Steady-State Performance with Dynamic Inductor Control

The steady-state performance of the dynamic inductor control concept is assessed through three tests described in the following subsections. The power supply Rohde and Schwarz HMP 2020 is used as a source voltage with the programmable DC electronic load BK Precision 8500 to perform the steady-state tests. These tests aim to demonstrate the BBC performance with dynamic inductor control under different operating points. The input current i(t) is also measured to show the impact of the dynamic inductor control on the input current ripple r value. A third order polynomial fit is used for better representation in the figures of sub-sections 7.1.1, 7.1.2 and 7.1.3.

7.1.1 Steady-State Performance at Different Load Currents

In this test, the source voltage V_s and the output voltage V_o are kept constant at 3.0 V and 4.2 V, respectively. The measurements are performed for the converter operating at light and heavy loads (from 5 % to 100 % of maximum load of 4.0 A) with dynamic inductor control disabled and enabled. Figure 7.1 depicts an increment of greater than 6 % at light loads and 24 % at maximum load in the efficiency when dynamic inductor control is enabled.

The nested relationship between the inductance value and conduction losses shown in Eq. 2.9 to Eq. 2.12 explains the greater efficiency improvement at heavy loads. Figure 7.2 and Figure 7.3 show the increment in the inductance

© Springer Fachmedien Wiesbaden GmbH, part of Springer Nature 2019
O. Abu Mohareb, *Efficiency Enhanced DC-DC Converter Using Dynamic Inductor Control*, Wissenschaftliche Reihe Fahrzeugtechnik Universität Stuttgart, https://doi.org/10.1007/978-3-658-25147-5_7

value when the DIC is enabled and the corresponding reduction in the ripple value, respectively. This increment in the inductance value leads to decrease the RMS input current, shown in Figure 7.4, and the conduction losses according to Eq. 2.9. The decrement in the conduction losses is also encountered by reduction in duty cycle D in order to restore the rated output current again, as shown in Figure 7.5. The decrement in the duty cycle additionally decreases the equivalent conduction resistance and voltage drop losses as shown in Figure 7.6 and Figure 7.7, respectively.

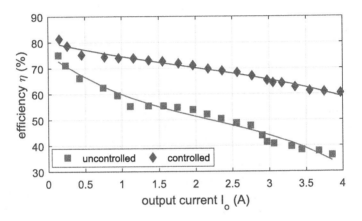

Figure 7.1:   Steady-state efficiency at different output (load) currents

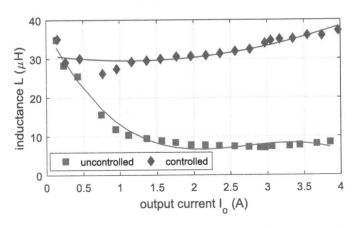

Figure 7.2:   Steady-state inductance values at different output (load) currents

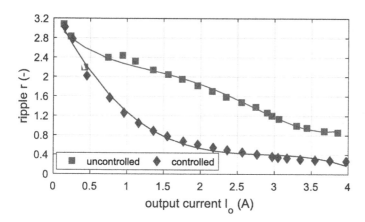

Figure 7.3: Steady-state ripple values at different output (load) currents

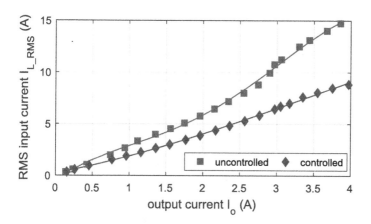

Figure 7.4: Steady-state RMS input current at different output (load) currents

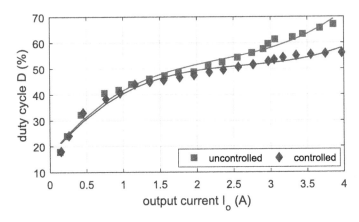

Figure 7.5:   Steady-state BBC duty cycle at different output (load) currents

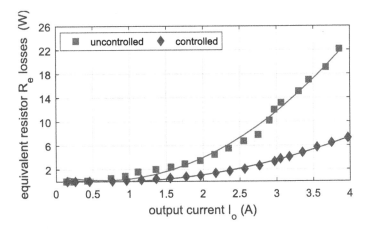

Figure 7.6:   Steady-state equivalent resistive losses at different output (load) currents

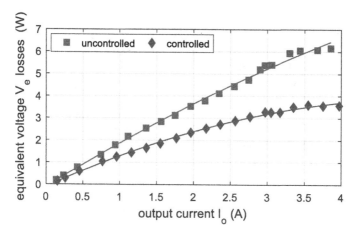

Figure 7.7: Steady-state equivalent voltage drop losses at different output (load) currents

7.1.2 Steady-State Performance at Different Source Voltages

This test aims to evaluate the DC-DC converter tolerance improvement when the DIC is used. During the test, the output current I_0 and the output voltage V_0 are kept constant at 2.0 A and 4.2 V, respectively. The measurements are performed for the converter operating at ±20 % of its nominal source voltage of 3.0 V with the above mentioned nominal output current and voltage when dynamic inductor control disabled and enabled. Enabling the DIC improves the ability of the DC-DC converter to tolerate the change in the source voltage, as depicted in Figure 7.8. The efficiency throw is only 8.86 % when the DIC is enabled and less than 56% of efficiency throw in the conventional DC-DC converter.

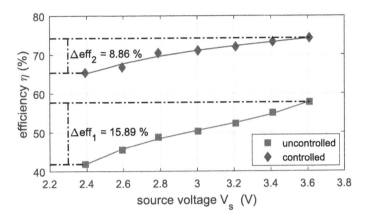

Figure 7.8:   Steady-state efficiency at different source voltages

7.1.3   Comparison of Efficiency with Conventional Boost Converter

Introducing the DIC to the DC-DC converter implements additional losses and suppresses the overall efficiency improvement. The simple loss model for the DIC consists of switching and conduction losses of the MOSFET and the free-wheeling diode. The DIC losses P_{ctr} are calculated as in [41] and shown in Figure 7.9. The DIC switching losses are negligible compared with the DIC conduction losses and the DC-DC converter losses, as depicted in Figure 7.9.

If the overall losses of a DC-DC converter with DIC P_{loss_all} (shown in Figure 7.10) are the sum of the DC-DC converter losses P_{loss} and the DIC losses P_{ctr}, the overall efficiency η_{all} can be calculated as in Eq. 7.1 and shown in Figure 7.11.

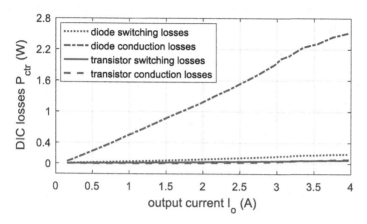

Figure 7.9:   DIC losses at different output (load) currents

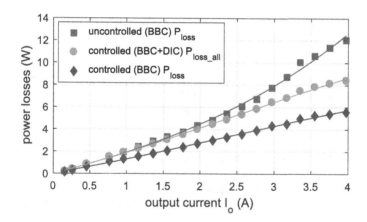

Figure 7.10:   Steady-state power losses at different output (load) currents

$$\eta_{all} = \frac{V_o\,I_o}{V_s\,I_{av} + P_{ctr}} = \frac{V_o\,I_o}{V_o\,I_o + P_{loss} + P_{ctr}} = \frac{V_o\,I_o}{V_o\,I_o + P_{loss\_all}} \qquad \text{Eq. 7.1}$$

Although Figure 7.11 depicts that improvement in the DC-DC converter efficiency is not affected by the turns ratio $N_2{:}N_1$, the overall efficiency is proportional to the turns ratio, where the required control current I_{ctr} is lower at higher turns ratio.

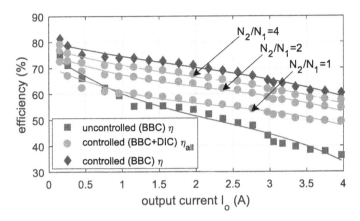

Figure 7.11:   Steady-state efficiency including DIC at different output (load) currents

The DIC losses can be minimal compared to the overall improvement in the efficiency through using a non-unity turns ratio of N_2:N_1 with higher turns number on the secondary winding to reduce the required control current.

7.1.4   Performance as Li-ion Battery Charger

Two types of control can be distinguished in the Li-ion battery charger; control of energy conversion and control of charging process. The energy control process is achieved by controlling duty cycle, while the control of charging process is implemented in the form of charging algorithm. A charging algorithm is a battery management function with which the battery itself is monitored and the energy conversion process in the charger is controlled in order to charge the battery in an efficient way [46].

The charge algorithm for Li-ion battery chemistries is a constant current or constant voltage (CC/CV) algorithm. The CC/CV charging algorithm described below is as stated in [81]. The charging cycle can be broken into two main stages: constant current charging and constant voltage charging, as shown in Figure 7.12. An additional stage called slow constant current charging may be required, where the battery is charged with a constant low charging current (typically 0.1C), if the battery is deeply depleted.

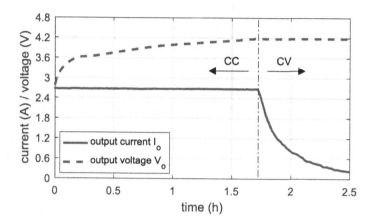

Figure 7.12:   Typical CC/CV charging profile for Li-ion battery

The battery is first charged with a constant current of a range, typically between 0.2C to 1.0C. The constant current charging stage ends when the cell voltage reaches its maximum voltage defined by the battery datasheet. At that point, the constant voltage charging begins. During this stage, the battery is charged with constant voltage and the charging current will start to fall due to internal cell resistance. The charging process is then terminated when the charging current falls below certain current limit.

The DC-DC converter has variable operating points during the charging process due to the continuous changing in the battery voltage and the charging current in the constant current and constant voltage stages, respectively. This test aims to evaluate the DC-DC converter performance as a Li-ion battery charger and shows the improvements when it is used with DIC. The PLB9744128VScLP rechargeable Li-ion polymer battery has replaced the programmable DC electronic load during this test. The charging parameters are set according to the product specifications of the Li-ion battery in [70], as listed in Table 7.1.

Table 7.1:   Charging parameters of the Li-ion battery

Parameter	Value	Note
Nominal voltage	3.7 V	-
Nominal capacity	5.4 Ah	-
Nominal energy	20 Wh	-
Voltage range	2.8 – 4.2 V	-
Nominal charging current	2.7 A	0.5C
Charging termination current	< 0.2 A	< 0.04C

During the test, the output current I_o is kept constant at 2.7 A in the CC mode and the output voltage V_o are kept constant at 4.2 V in the CV mode. The source voltage is set to 3.0 V and allowed to maximum fluctuate by ±3 % to emulate the real source voltage behavior, as shown in Figure 7.13. The battery is charged with the above mentioned charging parameters when DIC disabled and enabled. The DC-DC converter operating points vary between heavy (11.3 W) and light (0.83 W) loads, as shown in Figure 7.14.

The DIC is successfully able to follow the changes occurred during the charging process. The DIC algorithm continuously adapts the control current to follow the load change and achieve the maximum possible efficiency at all operating points, as shown in Figure 7.14. Enabling the DIC has remarkably improved the efficiency throw at heavy loads (CC mode) from 16.55 % at to be only around 1 %, as depicted in Figure 7.15. In addition, using the DIC guarantees higher DC-DC converter efficiency throughout the whole charging process.

Figure 7.16 shows the energy consumptions during the charging process. The DC-DC converter input charging energy is 44.28 Wh and 30.03 Wh when DIC disabled and enabled, respectively. Enabling the DIC consumes additional 3.31 Wh and the system total input energy will be 33.34 Wh, which is still much lower than the consumed input energy of the conventional DC-DC converter. Using the proposed DIC has saved 15 % of the energy required to fully-charge the Li-ion battery. Such great energy-saving amount is of much importance in battery-powered applications, especially in EV applications.

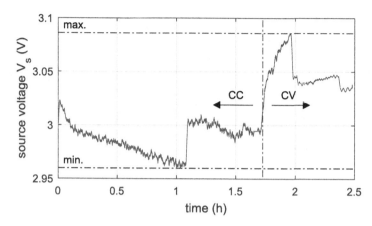

Figure 7.13:   Source voltage during charging Li-ion battery

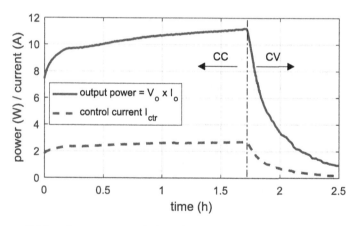

Figure 7.14:   Output power during charging Li-ion battery

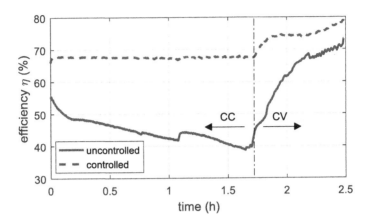

Figure 7.15:   Efficiency during charging Li-ion battery

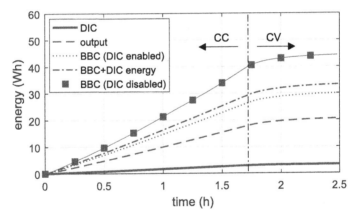

Figure 7.16:   Energy during charging Li-ion battery

7.2   Transient Performance of Converter with Dynamic Inductor Control

The transient performance of the DC-DC converter with the dynamic inductance control is appraised to demonstrate experimentally the effectiveness and stability of the proposed controller.

7.2.1   DC-DC Converter Transient Performance as DIC is Enabled

In this test, the source voltage V_s, the output voltage V_o and the output current I_o are kept constant at 3.0 V, 4.2 V and 2.7 A, respectively. The DC-DC converter is allowed to reach steady-state at first, and then the DIC is enabled as shown in Figure 7.17. After enabling the DIC, the converter reaches its steady-state again after approximately 180 ms, as shown in Figure 7.18, where the maximum inductance value and maximum efficiency for this operating point are achieved. The average input current I_{av} reaches its steady-state value of 5.98 A and remains stable. Figure 7.19 shows that the average control current I_{ctr} tends towards the minimum value of the average input current I_{av}. The higher inductance value resulting from enabling the DIC leads to lower ripple factor by more than 70 % (as in Figure 7.17) and thus a higher efficiency by about 19 %.

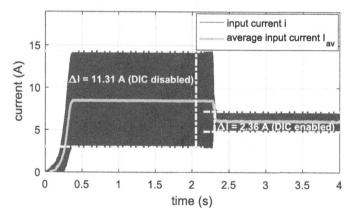

Figure 7.17:   Transients of converter input current (and its average) as DIC is enabled

The shaded area in Figure 7.18 shows that the DIC is enabled at time 2.274 s and the average input current almost reaches steady-state of 6.13 A after 50 ms at time 2.324 s. The DIC keeps increasing control current for the next 130 ms until the unique minimum average input current of 5.98 A with an average control current of 2.9 A at time 2.455 s is reached. Finding the unique minimum input current can further improve the efficiency from 61.66 % at time 2.324 s to be 63.21 % at time 2.455 s. After steady-state is reached, the average control current shown in Figure 7.19 keeps oscillating around its steady-state value according to the control method explained in section 4.4. The experimental results cope well with simulation results in section 0. Although the minor differences between the simulation and experiment due to the limits of the simulation models, it can be seen from both results that the proposed DIC is stable and effective at improving the DC-DC converter efficiency.

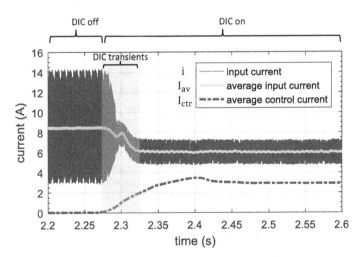

Figure 7.18: Transients converter input current (and its average) and the DIC control current

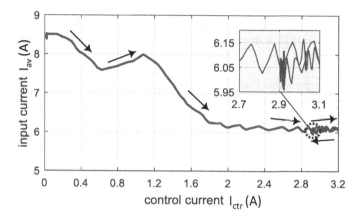

Figure 7.19:   Average input current as function of average control current for the prototype

It is worth mentioning that the duty cycle values in both PI controller for the DC-DC converter and the DIC are not allowed to change with big steps toward its final value in order to avoid undesirable sudden changes in the system. Together with the effect of the pre-charged output capacitor explained earlier in section 5.5, this explains the relative longer steady-state time of both the DC-DC controller and the DIC and the absence of the underdamped behavior of the DC-DC converter at the beginning and before enabling the DIC.

7.2.2   Step-Changed Source Voltage Transient with Enabled DIC

The wide spread of the photovoltaic and battery-powered technologies has forced the development of wide-range input DC-DC converters, as such sources have continuous-changing source voltage values. This test here shows the effectiveness and stability of the proposed controller with step-changed source voltage.

During the test, the output voltage V_o and the output (load) current I_o are kept constant at 3.7 V and 2.7 A, respectively. The under-sampled measurements shown in Figure 7.20 and Figure 7.21 are performed for the converter operating at a step varying source voltage ±6.67 % from 3.0 V with dynamic inductor control disabled and enabled, respectively. The programmable DC electronic load is used to step-change the source voltage. Due to its fast slew rate of 0.5 V/μs [82], it does not have a significant impact on the converter's step response.

Figure 7.21 depicts that the DC-DC converter with the DIC is able to handle the step change in source voltage without affecting system dynamics. Table 7.2 shows that the efficiency is not significantly affected by the source voltage change when the DIC is enabled. A voltage source change of 13.34 % of the rated voltage of 3.0 V has caused an efficiency throw of 13.62 % and 4.76 % when the DIC is disabled and enabled, respectively. The proposed controller has not only improved the efficiency at different source voltage values, but it has also effectively reduced the system sensitivity toward changing the source voltage.

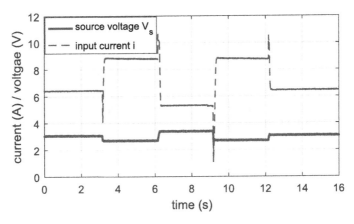

Figure 7.20: Input current during step-changed source voltage with DIC disabled

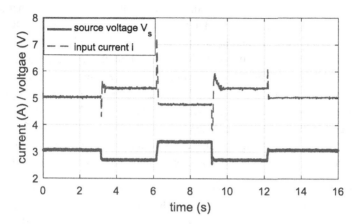

Figure 7.21:   Input current during step-changed source voltage with DIC enabled

Table 7.2:   Efficiency at different source voltages

| Source voltage (V) | Efficiency (%) | |
	DIC disabled	DIC enabled
2.8	43.07	64.92
3.0	51.20	67.51
3.2	56.69	69.68

7.2.3   Load Transient with Enabled DIC

In practice, many applications have a varying load at their outputs. A good example of continuously-varying load is the Li-ion battery charger, where the output power is varying all the time during the charging process. Accordingly, the proposed DC-DC converter with DIC should be able to handle the load continuous varying behavior. Sub-section 7.1.4 has shown the ability of the DC-DC converter with DIC to cope very well with continuous change of the load and its output power, as it keeps working at its maximum possible efficiency all the time.

Nevertheless, the change in the output power during the Li-ion charging process is relatively slow. In order to evaluate the ability of the DC-DC converter with DIC to handle more aggressive load changes, the load transient performance of the DIC is appraised in the presence of a step varying load. During the test, the source voltage V_s and the output (load) current I_o are kept constant at 3.0 V and 2.7 A, respectively. The under-sampled measurements shown in Figure 7.22 and Figure 7.23 are performed for the converter operating at a step varying output voltage from 3.7 V to 4.2 V with dynamic inductor control disabled and enabled, respectively. The programmable DC electronic load is used to step-change the output load.

The DC-DC converter reaches steady-state in about 95 ms and 105 ms with DIC disabled and enabled, respectively. The DC-DC converter with the DIC enabled is slightly slower than with the DIC disabled. Enabling the DIC forces a higher inductance value which tends to damp the converter's response, as depicted in Eq. 3.34.

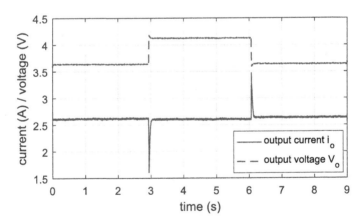

Figure 7.22: Output voltage and current during load transient with DIC disabled

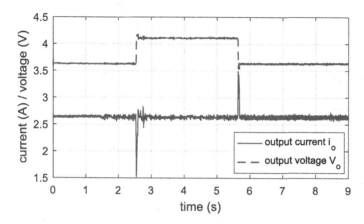

Figure 7.23: Output voltage and current during load transient with DIC enabled

Based on load transient measurements and using Eq. 2.12, Eq. 3.34, Eq. 3.35, Eq. 3.39 and Eq. 3.42, Table 7.3 compares the dynamics of the DC-DC converter when DIC is disabled and enabled. The DC-DC converter operates in the CCM, as the inductance value is greater than CCM inductance. An underdamped behavior is observed from Figure 7.22 and Figure 7.23, as the damping ratio is well below 1 in both cases.

The DIC is clearly able to provide a better energy transfer due to its higher inductance and without losing systems dynamics as the inductance is below the MET inductance value. Accordingly, the DIC provides higher efficiency without undesired interactions with the converter's controller.

Table 7.3:   Load transient parameters with DIC disabled and enabled

Parameter and unit	Symbol	DIC disabled	DIC enabled	Note
Source voltage (V)	V_s	3.00	3.00	-
Output voltage (V)	V_o	3.70	3.70	-
Average input current (A)	I_{av}	6.74	4.85	-
Output current (A)	I_o	2.70	2.70	-
Input current peak-to-peak (A)	ΔI	11.12	3.37	-
Duty cycle (-)	D	0.47	0.44	-
Inductance (μH)	L	6.34	19.59	Eq. 2.12
Damping ratio (-)	ζ	0.021	0.035	Eq. 3.34 at 6600 μF
Critically damped inductance (μH)	L_c	13926	15547	Eq. 3.35 at 6600 μF
CCM inductance (μH)	L_{CCM}	4.52	4.73	Eq. 3.39
MET inductance (μH)	L_{MET}	43.20	46.14	Eq. 3.42
Efficiency without DIC losses (%)	η	48.33	67.52	-

The formerly shown steady-state and the transient tests have emphasized the effectiveness of the novel DIC in improving the DC-DC converter efficiency at various operating points and without losing the system dynamics.

8 Conclusions and Future Work

Switched-mode DC-DC converters are the essential part of much electronic equipment for their high efficiency and simplicity. Conventional DC-DC converters suffer from efficiency degradation due to continuous change in the inductance value during operation. This dissertation has proposed a novel dynamic inductor control that can be generally applied to various DC-DC converter types to improve the converter efficiency throughout controlling the inductance value at all operating points without consequential complexity and increase in the inductor cost and size.

The major contributions of this dissertation include the design of the variable inductor structure, the development of dynamic inductor control concept, the conduction of simulation models and a prototype to implement the proposed inductor structure and control concept with a conventional DC-DC converter and perform simulation and experimental validation of the dynamic inductor control.

8.1 Conclusions

The work in this dissertation has introduced the dynamic inductor control concept to improve the DC-DC converter efficiency. The proposed concept provides improved efficiency throughout maintaining the inductance at its maximum possible value during operation, and thus, reducing the conduction and RMS losses at all operation points.

The present work was designed to determine the effect of the variation in the inductance value on the converter efficiency and develop a control concept that overcomes the limitations of the conventional inductors. The investigation has shown that the inductance value is directly affected by the current passing through it and the core tends to saturate when high currents are passing through resulting in a lower inductance value. The proposed dynamic inductor control aims to operate the inductor at magnetic field intensity where the core permeability and the inductance are at their maximum values.

© Springer Fachmedien Wiesbaden GmbH, part of Springer Nature 2019
O. Abu Mohareb, *Efficiency Enhanced DC-DC Converter Using Dynamic Inductor Control*, Wissenschaftliche Reihe Fahrzeugtechnik Universität Stuttgart, https://doi.org/10.1007/978-3-658-25147-5_8

The development of a non-ideal state-space averaging model for DC-DC boost converter in this work allowed detailed analysis and modeling of the non-ideal boost converter interfaced to non-ideal DC power sources and non-ideal battery load. The detailed model has enabled defining the maximum energy transfer concept. This concept has defined the limits of inductance value where the conduction and RMS losses are minimized and the system response is kept fast enough for the load changes through the whole operating range. The investigation has shown that inductance values beyond the maximum inductance value defined by maximum energy transfer concept has a negligible improvement of the efficiency and tends to deteriorate system dynamics against load changes.

Simulation and experimental results confirm the effectiveness and the high capability of the proposed concept in improving efficiency of various DC-DC converter topologies at all operating points. The main advantages of the dynamic inductor control compared to the prior art of inductor control in DC-DC converters are summarized in the following:

- The output values of the converter are regulated by the mean of the conventional duty cycle control and not using the proposed control concept. This makes the dynamic inductor control applicable to the conventional converter designs without major circuitry changes.

- The proposed dynamic inductor control (DIC) is only meant for improving the efficiency at any desired operating point, including both light and heavy loads.

- The proposed control method avoids the contradictions between regulating the output values and having the highest possible efficiency, as well as, it preserves fast systems dynamics against load changes.

- No complex specially-designed inductor is required and the proposed variable inductor structure can be easily adapted to any conventional DC-DC converter.

- The control function between the primary and control flux is independent from the physical core shape and geometry and directly related to the turn ratio between the primary and secondary windings.

■ The proposed dynamic control is the first of its kind that takes the efficiency of the DC-DC converter as a control parameter without limiting its output power or using bigger inductor.

It may be concluded that the proposed dynamic inductor control offers greatly improved efficiency and it has merit where these advantages justify the extra needed control circuitry.

8.2 Future Work

The work in this dissertation has thrown up many questions in need of further investigation. It is recommended that further research should be undertaken in the following aspects:

■ The maximum energy transfer concept is manually proved for different operating points. Since the maximum energy transfer inductance value varies with different loads or input voltage values, future research should therefore concentrate on the investigation of automatic recognition of maximum energy transfer inductance with the dynamic inductor control.

■ An important limitation of the dynamic inductor control circuitry in Figure 6.4 needs to be considered. The implemented circuitry in this work assumes that all operating points have magnetic field intensity higher than that at the inductor maximum permeability point and an opposing flux is only required to get the maximum possible inductance value. Therefore, a modified dynamic inductor control circuitry design should be in the future considered, where the flux generated for the control current may aid or oppose the load flux. This also implies investigating a modified control methodology that is capable of supporting both cases.

■ A commonly used toroid ferrite cores inductor was chosen in this work to implement and evaluate the proposed dynamic inductor control concept. Ferrite cores for switched-mode DC-DC converters are made in a variety of shapes and sizes and it would be interesting to assess implementing the dynamic inductor control concept with other cores' shapes.

■   The ability of the proposed inductor control in improving the DC-DC converter efficiency at different loads or at different source voltages was proved throughout this dissertation. But in many applications where the source voltage at the input of the converter and the load at its output can simultaneously vary, such as renewable energy sources with storage systems, the effectiveness of DIC in improving the efficiency should be further investigated.

References

[1] A. I. Pressman, K. Billings und T. Morey, Switching power supply design, New York: McGraw-Hill, 2009.

[2] P. Midya und P. T. Krein, „Feed-forward active filter for output ripple cancellation," Bd. 77, Nr. 5, pp. 805-818, 1994.

[3] V. T. Liu und L. J. Zhang, „Design of high efficiency boost-forward-flyback converters with high voltage gain," in *11th IEEE International Conference on Control & Automation (ICCA)*, Taichung, Taiwan, 2014.

[4] M. Das und V. Agarwal, „A novel, high efficiency, high gain, front end DC-DC converter for low input voltage solar photovoltaic applications," in *38th annual conference on IEEE industrial electronics society*, Montreal, QC, Canada, 2012.

[5] Y. Cai, D. Xu, Z. Chen und S. Zhong, „Analysis and design of a high efficiency high step-up gain DC-DC converter," pp. 6-10, 2014.

[6] J. Zhang, „Bidirectional DC-DC power converter design optimization, modeling and control," Blacksburg, Virginia, USA, 2008.

[7] W.-Y. Choi, J.-S. Yoo, J.-Y. Choi, M.-K. Yang und H.-S. Cho, „High efficiency step-up DC-DC converter for low-DC renewable energy sources," pp. 1417-1421, 2012.

[8] S. M. A. Iqbal, „Buck and boost converter design optimization parameters in modern VLSI technology," in *International conference and seminar on micro/nanotechnologies and electron devices proceedings*, Erlagol, Altai, 2011.

[9] K. Morimoto, T. Doi, H. Manabe, M. Nakaoka, H.-W. Lee, N. A. Ahmed, E. Hiraki und T. Ahmed, „Next generation high efficiency high power dc-dc converter incorporating active switch and snubbing capacitor assisted full-bridge soft-switching PWM inverter with high

© Springer Fachmedien Wiesbaden GmbH, part of Springer Nature 2019
O. Abu Mohareb, *Efficiency Enhanced DC-DC Converter Using Dynamic Inductor Control*, Wissenschaftliche Reihe Fahrzeugtechnik Universität Stuttgart, https://doi.org/10.1007/978-3-658-25147-5

frequency transformer for large current output," in *20th annual IEEE applied power electronics conference and exposition*, 2005.

[10] M. Venkatesh Naik und P. Samuel, „Analysis of ripple current, power losses and high efficiency of DC-DC converters for fuel cell power generating systems," *Journal of renewable and sustainable energy*, pp. 1080-1088, 2016.

[11] H. Fathabadi, „Novel photovoltaic based battery charger including novel high efficiency step-up DC/DC converter and novel high accurate fast maximum power point tracking controller," pp. 200-211, 2015.

[12] W.-J. Cha, J.-M. Kwon und B.-H. Kwon, „Highly efficient step-up dc–dc converter for photovoltaic micro-inverter," pp. 14-21, 2016.

[13] Maxima Integrated, „DC-DC converter tutorial," San Jose, California, USA, Nov. 2001.

[14] „How to design an efficient DC-DC converter using the DS1875 PWM controller," San Jose, California, USA, October 2012.

[15] K. Frick, „Inductor choice yields performance tradeoffs in DC-DC converters," San Jose, California, US, Feb. 2007.

[16] G. Sizikov, A. Kolodny, E. G. Fridman und M. Zelikson, „Efficiency optimization of integrated DC-DC buck converters," in *17th IEEE International Conference on Electronics, Circuits, and Systems (ICECS)*, Athens, 2010.

[17] P. Mali, Magnetic amplifiers principles and applications, New York, USA: John F. Rider Publisher, 1960.

[18] R. Lee, Electronic transformers and circuits, New York, USA: John Wiley & Sons, 1955.

[19] L. W. Orr, „Permeability measurements in magnetic ferrites," Ann Arbor, Michigan, USA, September 1952.

[20] C. Vaucourt, „Choosing inductors and capacitors for DC/DC converters," Dallas, Texas, USA, February 2004.

[21] A. V. Bakshi, Electromagnetic field theory, India: Technical Publications Pune, 2009.

[22] R. A. Salas und J. Pleite, „Simple procedure to compute the inductance of a toroidal ferrite core from the linear to the saturation regions," June 2013.

[23] W. G. Hurley und W. H. Wölfle, Transformers and inductors for power electronics: theory, design and applications, UK: John Wiley & Sons, 2013.

[24] A. Raj, „Calculating efficiency," Dallas, Texas, USA, February 2010.

[25] R. Shaffer, Fundamentals of power electronics with MATLAB, Boston, USA: Charles River Media, 2007.

[26] G. Trinkaus, „Magnetic amplifiers: another lost technology," High Voltage Press, USA, 1951.

[27] R. L. Brandt, „DC-DC converter having magnetic feedback". USA/Orange, CA Patent US 7,378,828 B2, 27 May 2008.

[28] E. Haugs und F. Strand, „Magnetically controlled inductive device". USA/Sperrebotn, NO Patent US 7,256,678 B2, 14 August 2007.

[29] J. Kugler, A. Dvorak und C. Elliott, „Protection circuit for magnetically controlled D-C to D-C converter". Canada/Ottawa Patent US 4,005,352, 25 january 1977.

[30] A. L. Peterson und H. C. Martin, „Control circuit for a DC-DC power converter including a controlled magnetic core flux resetting technique for output regulation". USA/Escondido, CA Patent US 5,392,206, 21 February 1995.

[31] W. H. Coley, C. . E. Hawkes und K. D. Mathews, „DC/DC converter with magnetic flux density limits". USA/Cary, North Carolina Patent EP 2 393 190 A2, 7 December 2011.

[32] W. H. Coley, C. E. Hawkes und K. D. Mathews, „DC/DC converter with magnetic flux density limits". USA/Cary, North Carolina Patent US 8,693,215 B2, 8 April 2014.

[33] „Switch-mode power supply Reference Manual Rev. 4," ON Semiconductor, Phoenix, Arizona, USA, April 2014.

[34] G. A. Rincón-Mora und N. Keskar, „Unscrambling the power losses in switching boost converters," *EE Times,* 2006.

[35] B. T. Lynch, „Under the hood of a DC/DC boost converter," Texas Instruments, Dallas, Texas, USA, 2008/09.

[36] Q. A. Naman, O. Abu Mohareb und Q. Jaber, „A novel variable duty cycle half-forward converter having low current harmonics and high power factor," in *Power Electronics, Machines and Drives (PEMD 2010), 5th IET International Conference,* Brighton, UK, 2010.

[37] „Power factor correction (PFC) basics," Fairchild Semiconductor, San Jose, California, USA, August 2004.

[38] R. W. Erickson und D. Maksimović, Fundamentals of power electronics, New York: Kluwer Academic Publishers, 2001.

[39] X. Cheng und G.-j. Xie, „Full order models and simulation of boost converters operating in DCM," in *IEEE 2009 International Conference on Electronic Computer Technology,* Macau, 2009.

[40] J. R. Warren, K. A. Rosowski und D. J. Perreault, „Transistor selection and design of a VHF DC-DC power converter," *IEEE Transactions on Power Electronics vol. 23, no. 1,* pp. 27-37, January 2008.

[41] S. Jaunay und J. Brown, „DC-to-DC design guide," Vishay Siliconix, October 2002.

[42] N. Mohan, Power electronics: a first course, Minneapolis, MN, USA: John Wiley & Sons, 2012.

[43] O. Abu Mohareb, H.-C. Reuss, Q. A. Naman, M. Grimm und O. Badran, „Effect of dissipative elements on boost converter behaviour," in *PCIM Europe*, Nuremburg, Germany, 2011.

[44] S. A. Ilangovan und S. Sathyanarayana, „Impedance parameters of individual electrodes and internal resistance of sealed batteries by a new nondestructive technique," *Journal of Applied Electrochemistry*, pp. 456-563, 1992.

[45] S. C. Hageman, „Simple PSpice models let you simulate common battery types," *Electronic Design News*, pp. 117-132, 1993.

[46] H. J. Bergveld, Battery management systems design by modelling, Nederlands: Royal Philips Electronics N.V, 2001.

[47] P.-L. Huynh, O. Abu Mohareb, M. Grimm, H.-J. Mäurer, A. Richter und H.-C. Reuss, „Impact of cell replacement on the State-of-Health for parallel li-ion battery pack," in *IEEE Vehicle Power and Propulsion Conference (VPPC)*, Coimbra, Portugal, 2014.

[48] P.-L. Huynh, O. Abu Mohareb, H.-C. Reuss, M. Grimm, H.-J. Mäurer und A. Richter, „Einfluss der Architektur von Lithium-Ionen Akkumulatoren auf deren charakterisierende Parameter und deren Bestimmung," in *TAE Elektromobilität*, Ostfildern, Germany, 2014.

[49] S. Buller, Impedance-based simulation models for energy storage devices in advanced automotive power systems (Dissertation), Aachen, Germany: Shaker Verlag GmbH, 2003.

[50] P. Horowitz und W. Hill, The art of electronics, New York, USA: Cambridge University Press, 1989.

[51] M. K. Kazimierczuk, G. Sancineto, G. Grandi, U. Reggiani und A. Massarini, „High-frequency small-signal model of ferrite core inductors," *IEEE Transactions on Magnetics*, pp. 4185-4191, 1999.

[52] M. Bartoli, A. Reatti und M. K. Kazimierczuk, „High-frequency models of ferrite core inductors," in *20th International Conference on Industrial Electronics, Control and Instrumentation*, Bologna, 1994.

[53] J. Liu, T. G. Wilson, R. C. Wong, R. Wunderlich und F. C. Lee, „A method for inductor core loss estimation in power factor correction applications," in *Seventeenth Annual IEEE Applied Power Electronics Conference and Exposition*, Dallas, TX, USA, 2002.

[54] Maxim Integrated Products, Inc., „An efficiency primer for switch-mode, DC-DC converter power supplies," San Jose, California, USA, December 2008.

[55] G. M. Buiatti, A. M. R. Amaral und A. J. M. Cardoso, „ESR estimation method for DC/DC converters through simplified regression models," in *Industry Applications Conference. 42nd IAS annual meeting*, New Orleans, LA, USA, 2007.

[56] E. Taddy und V. Lazarescu, „Modeling and simulation of buck dc-dc converter with capacitor equivalent series resistance," in *6th international conference on Electronics, Computers and Artificial Intelligence (ECAI)*, Bucharest, Romania, 2014.

[57] N. D. Benavides und P. L. Chapman, „Mass-optimal design methodology for DC-DC converters in low-power portable fuel cell applications," *IEEE Transactions on Power Electronics*, pp. 1545-1555, 2008.

[58] R. D. Middlebrook und S. Cuk, „A general unified approach to modelling switching-converter power stages," in *IEEE Power Electronics Specialists Conference*, Cleveland, Ohio, USA, 1976.

[59] A. S. Kislovsk, R. Redl und N. O. Sokal, Dynamic analysis of switching-mode DC/DC converters, New York, USA: Van Nostrand Reinhold, 1991.

[60] S. M. Ćuk, „Modelling, analysis, and design of switching convert-ers," California Institute of Technology, Pasadena, California, USA, 1977.

[61] C. Zhou, R. B. Ridley und F. C. Lee, „Design and analysis of a hysteretic boost power factor correction circuit,“ in *21st Annual IEEE Conference on Power Electronics Specialists*, San Antonio, TX, USA, 1990.

[62] EPCOS AG, „Power line chokes: current-compensated ring core double chokes datasheet,“ 2012.

[63] L. H. Dixon, „Magnetics design for switching power supplies,“ Dallas, Texas, USA, 2001.

[64] R. Balog und P. T. Krein, „Automatic tuning of coupled inductor filters,“ in *IEEE 33rd Annual Power Electronics Specialists Conference*, Cairns, Queensland, Australia, 2002.

[65] D. S. Lymar, „Coupled-magnetic filters with adaptive inductance cancellation,“ Massachusetts Institute of Technology, Cambridge, Massachusetts, USA, June 2005.

[66] D. L. Logue und P. T. Krein, „Optimization of power electronic systems using ripple correlation control: a dynamic programming approach,“ in *IEEE 32nd Annual Power Electronics Specialists Conference*, Vancouver, BC, 2001.

[67] S.-J. Liu und M. Krstic, Introduction to extremum seeking, London: Springer, 2012.

[68] D. G. Luenberger, Introduction to dynamic systems: theory, models, and applications, New York, USA: John Wiley & Sons, 1979.

[69] S. Chapra und R. Canale, Numerical methods for engineers, New York: McGraw-Hill, 2010.

[70] J. Cao und Y. Choi, „PLB9744128VScLP rechargeable lithium ion polymer battery (Product specification),“ South Korea, 2012.

[71] P.-L. Huynh, Beitrag zur Bewertung des Gesundheitszustands von Traktionsbatterien in Elektrofahrzeugen, Stuttgart, Germany: Springer Vieweg, 2016.

[72] D. Andre, M. Meiler, K. Steiner, C. Wimmer, T. Soczka-Guth und D. U. Sauer, „Characterization of high-power lithium-ion batteries by electrochemical impedance spectroscopy. I. Experimental investigation," *Journal of Power Sources,* pp. 5334-5341, 2011.

[73] W. Waag, S. Käbitz und D. U. Sauer, „Experimental investigation of the lithium-ion battery impedance characteristic at various conditions and aging states and its influence on the application," *Applied Energy,* pp. 885-897, 2012.

[74] B. Hauke, „Basic calculation of a buck converter's power stage," Texas Instruments, Dallas, Texas, USA, December 2011.

[75] T.-T. Tay, I. Mareels und J. B. Moore, High performance control, Boston, USA: Birkhauser Boston Inc., 1997.

[76] M. Hagen und V. Yousefzadeh, „Applying digital technology to PWM control-loop designs," in *TI Power Supply Design Seminar series,* Dallas, Texas, USA, 2008.

[77] J. J. Carr und J. M. Brown, Introduction to biomedical equipment technology, Upper Saddle River, New Jersey, USA: Prentice Hall, 1998.

[78] J. Vetelino und A. Reghu, Introduction to sensors, Boca Raton, Florida, USA: CRC Press, 2011, p. 17.

[79] F.W. Bell, „NT series magneto-resistive current sensors," Orlando, Florida, USA, 04.2003.

[80] H. Lutz und W. Wendt, Taschenbuch der Regelungstechnik mit MATLAB und Simulink, Frankfurt am Main, Germany: Harri Deutsch, 2010, p. p. 81.

[81] H. Grewal, „Li-ion battery charger solution using the MSP430," Texas Instruments, Dallas, Texas, USA, December 2005.

[82] B&K Precision, „B&K Precision 8500 series DC electronic loads," B&K Precision, Yorba Linda, Canada, 2016.

Appendix

A1.   Battery Boost Charger Canonical Model

The first step in obtaining a canonical model is to derive the small-signal AC equivalent circuit. The small-signal model of CCM boost converter is derived by averaging the switching ripples in the inductor current and capacitor voltage waveforms during one switching cycle. Hence, using Eq. 3.13 and Eq. 3.14, the DC and low-frequency AC variations in the inductor voltage and the capacitor current are as in Eq. A.1 and Eq. A.2 respectively, where $\langle d'(t)\rangle = 1 - \langle d(t)\rangle$ and $\langle d(t)\rangle$ are the DC and low-frequency AC variations in the duty cycle.

$$
\begin{aligned}
\langle v_L(t)\rangle &= L\frac{d\langle i(t)\rangle}{dt} \\
&= \langle v_s(t)\rangle - \langle d(t)\rangle V_T - \langle d'(t)\rangle V_D \\
&\quad - \langle d'(t)\rangle \frac{R_c}{R_b + R_c}\langle v_b(t)\rangle \\
&\quad - \langle d'(t)\rangle \frac{R_b}{R_b + R_c}\langle v_c(t)\rangle - (R_s + R_L)\langle i(t)\rangle \\
&\quad - \langle d(t)\rangle R_{on}\langle i(t)\rangle - \langle d'(t)\rangle R_D\langle i(t)\rangle \\
&\quad - \langle d'(t)\rangle \frac{R_b R_c}{R_b + R_c}\langle i(t)\rangle
\end{aligned}
\qquad \text{Eq. A.1}
$$

$$
\langle i_c(t)\rangle = C\frac{d\langle v_c(t)\rangle}{dt} = \langle d'(t)\rangle \frac{R_b}{R_b + R_c}\langle i(t)\rangle - \frac{\langle v_c(t)\rangle - \langle v_b(t)\rangle}{R_b + R_c}
\qquad \text{Eq. A.2}
$$

Assume that the input source voltage $v_s(t)$, the battery voltage $v_b(t)$ and the duty cycle $d(t)$ are equal to some given quiescent values at steady-state plus some superimposed small AC variations as in Eq. A.3. In this equation, the battery voltage $v_b(t)$ is introduced as a new input for the non-ideal BBC.

© Springer Fachmedien Wiesbaden GmbH, part of Springer Nature 2019
O. Abu Mohareb, *Efficiency Enhanced DC-DC Converter Using Dynamic Inductor Control*, Wissenschaftliche Reihe Fahrzeugtechnik Universität Stuttgart, https://doi.org/10.1007/978-3-658-25147-5

$$\left.\begin{array}{l} \langle v_s(t)\rangle = V_s + \hat{v}_s(t) \\ \langle v_b(t)\rangle = V_b + \hat{v}_b(t) \\ \langle d(t)\rangle = D + \hat{d}(t) \\ \langle d'(t)\rangle = D' - \hat{d}(t) \end{array}\right\} \Rightarrow \begin{array}{l} \langle v_c(t)\rangle = V_c + \hat{v}_c(t) \\ \langle i(t)\rangle = I + \hat{i}(t) \end{array} \qquad \text{Eq. A.3}$$

Hence, a small-signal linearized model with first-order AC terms around these quiescent operating points can be obtained by substituting the terms in Eq. A.3 into Eq. A.1 and Eq. A.2, then separating the DC and AC terms. Each of the resultant terms contains a single AC quantity usually multiplied by a constant coefficient such as a DC term and they are the linear functions of the AC variations, as in Eq. A.4, Eq. A.5 and Eq. A.6. These first-order AC terms represent the voltage across the inductor and current through the capacitor in the small-signal model.

$$\begin{aligned} L\frac{d\,\hat{i}(t)}{dt} = \hat{v}_s(t) - \hat{i}(t)\Big(&R_s + R_L + DR_{on} + (1-D)R_D \\ &+ (1-D)\frac{R_b R_c}{R_b + R_c}\Big) \\ &+ \hat{d}(t)\left(-V_T + V_D + \frac{R_c}{R_b + R_c}V_b + \frac{R_b}{R_b + R_c}V_c\right) \\ &+ I\left(-R_{on} + R_D + \frac{R_b R_c}{R_b + R_c}\right)\Big) \\ &- \left(\frac{(1-D)}{R_b + R_c}\right)(\hat{v}_b(t)R_c + \hat{v}_c(t)R_b) \end{aligned} \qquad \text{Eq. A.4}$$

$$C\frac{d\,\hat{v}_c(t)}{dt} = \hat{i}(t)(1-D)\frac{R_b}{R_b + R_c} - \hat{d}\frac{R_b}{R_b + R_c}I + \frac{\hat{v}_b(t) - \hat{v}_c(t)}{R_b + R_c} \qquad \text{Eq. A.5}$$

$$\hat{v}_o(t) = \hat{v}_b(t) + \hat{i}_o(t)R_b = \hat{v}_c(t) + \hat{i}_c(t)R_c \qquad \text{Eq. A.6}$$

The terms of Eq. A.4 and Eq. A.5 are equal to other voltage and current dependent sources. Accordingly, they are used to construct two equivalent circuits around the inductor voltage and the capacitor current, as shown in Figure A.1, where V_{ee} and R_{ee} are defined in Table 3.1. Theses equivalent circuits can be combined into one small-signal AC equivalent circuit via AC transformer of a ratio similar to that in the DC transformer, as shown in Figure A.2. By manipulating Figure A.2 using Thévenin/Norton theorem, the canonical model is obtained, as shown in Figure 3.15.

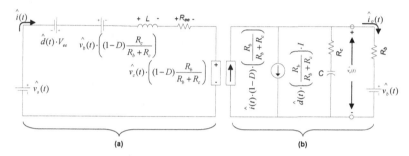

Figure A.1: Non-ideal BBC equivalent circuits corresponding to Eq. A.4 and Eq. A.5

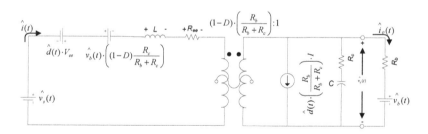

Figure A.2: Non-ideal BBC small-signal AC equivalent circuit

A2. Simulink Models

This section includes the complete BBC with control strategy of Figure A.3 and the details for each Simulink block. The Matlab code for dynamic inductor control algorithm of section 4.3 is given.

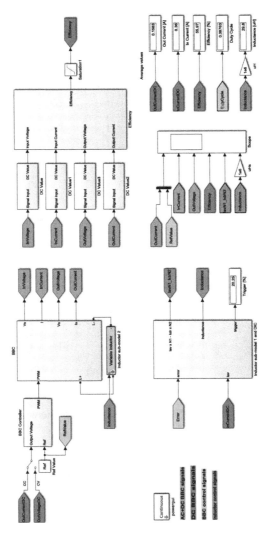

Figure A.3: Complete BBC Simulink model with control strategy

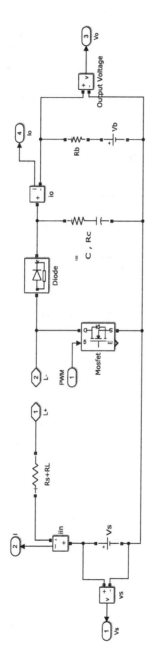

Figure A.4: BBC Simulink model

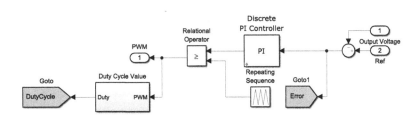

Figure A.5: BBC PI controller and PWM generator

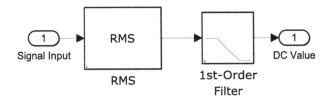

Figure A.6: Simulink block for RMS value with low-pass filter

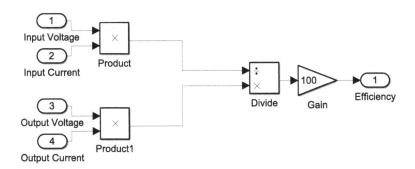

Figure A.7: Simulink block for calculating the efficiency

Matlab code for dynamic inductor control algorithm in Figure 4.8:

```
function[ICTR, on_trig] = fcn(Iinr)

persistent Iin Imin N2N1 B N dIctr dImin fst Lcrt_firt
kk Ictr

if isempty(fst)
    fst = 1;
    Lcrt_firt = 15e2;%7.60358e7;
    kk = 0;
    Ictr = 0;

    Imin = 0;
end

Iin = Iinr;
N2N1  = 1;
B     = 0.48;
N     = N2N1 / B;

dIctr = 0.001;
dImin = 0.003;

% call L control every 15 times ones
if kk > Lcrt_firt
    lmt = 0.0035;
    if Ictr > 0 % 6
        dlt = Iin/N - Imin;
        if dlt > 0 && dlt < lmt % 8
            Ictr = Ictr - dIctr; % 9
            if Iin/N <= Imin % 10
                Imin = Iin/N; % 11 (check)
            end
        else
            Ictr = Ictr + 2*dIctr; % 12
            Imin = Imin + dImin; %13
        end
    end
    if Ictr <= 0 % 0
        Ictr = Iin/N; % 1
        Iminp = Iin/N; % 2
    end
```

```
    Lcrt_firt = 2000; % call L control every 15 times
ones
    kk = 0;
end

kk = kk + 1;

% keep running the control current (first time is set
to zero)
if Ictr > Iin/N % 3 (yes)
        Ictr = Iin/N; % 4
        Imin = Iin/N + dImin; % 5

end

ICTR = Ictr;
on_trig = (kk/Lcrt_firt)*100;
end
```

A3.   Schematic and Bill of Materials for the Prototype

This section presents the schematic and the bill of materials for the prototype converter with dynamic inductor control of chapter 6.

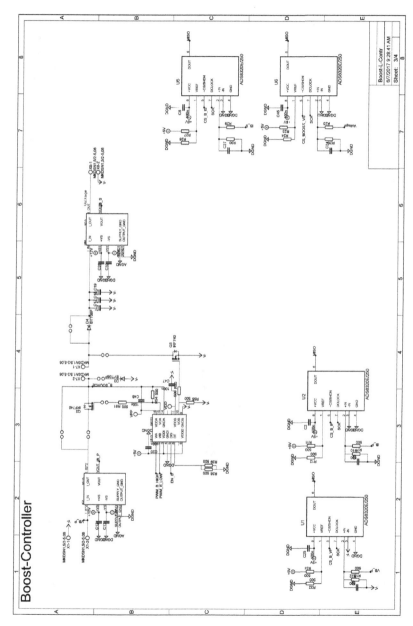

Figure A.8: Schematic of the BBC with anlog to digital converters

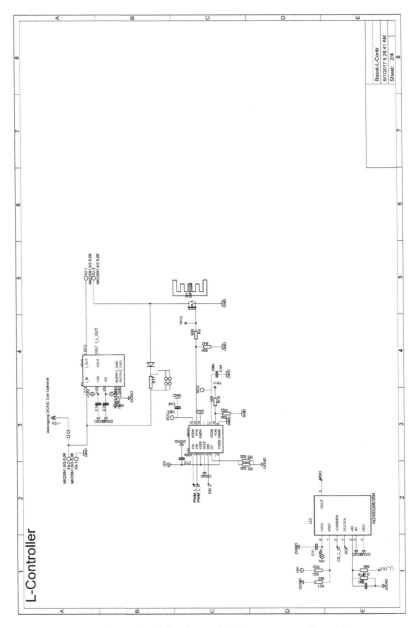

Figure A.9: Schematic of the dynamic inductor control circuitry

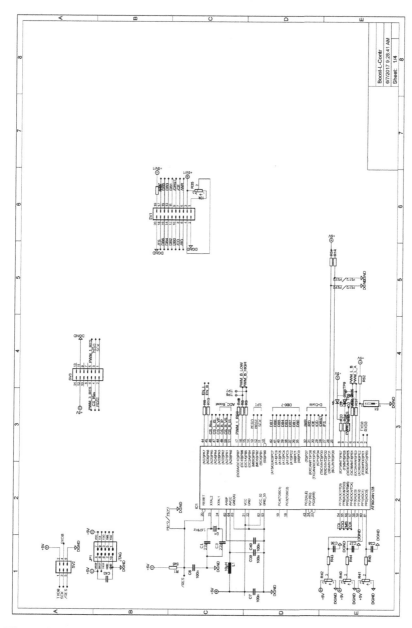

Figure A.10: Schematic of the BBC and DIC control unit

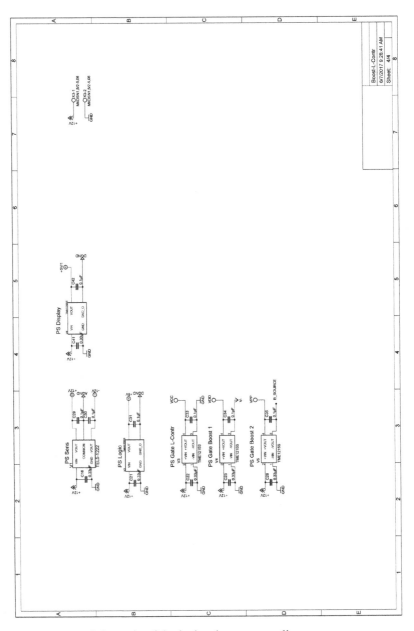

Figure A.11: Schematic of the isolated power supplies

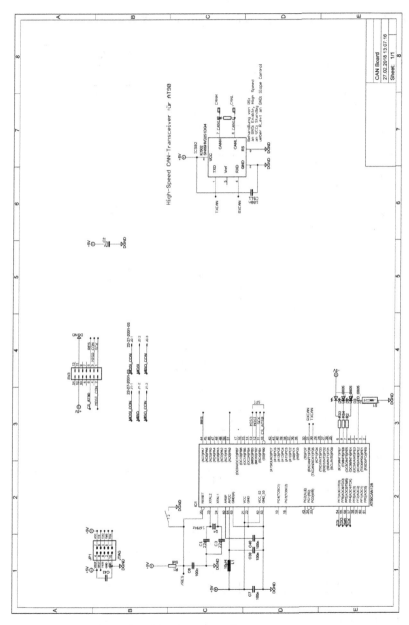

Figure A.12: Schematic of the CAN board

Table A.1:   Bill of materials for the prototype PCB

Part	Value	Device	Package
C1	22p	C-EUC0805	C0805
C2	100n	C-EUC0805	C0805
C3	22p	C-EUC0805	C0805
C4	C-EUC0805	C0805	rcl
C5	C-EUC0805	C0805	rcl
C6	100n	C-EUC0805	C0805
C7	100n	C-EUC0805	C0805
C8	100n	C-EUC0805	C0805
C9	C-EUC0805	C0805	rcl
C10	100n	C-EUC0805	C0805
C11	100n	C-EUC0805	C0805
C12	100n	C-EUC0805	C0805
C13	100n	C-EUC0805	C0805
C14	100n	C-EUC0805	C0805
C15	100n	C-EUC0805	C0805
C16	0.33uF	C-EUC0805	C0805
C17	CPOL-EUE5-13	E5-13	rcl
C18	CPOL-EUE5-13	E5-13	rcl
C19	CPOL-EUE5-13	E5-13	rcl
C20	100n	C-EUC0805	C0805
C21	0.33uF	C-EUC0805	C0805
C22	0.33uF	C-EUC0805	C0805
C23	0.33uF	C-EUC0805	C0805
C24	C-EUC0805	C0805	rcl
C25	100n	C-EUC0805	C0805
C26	100n	C-EUC0805	C0805
C27	C-EUC0805	C0805	rcl
C28	0.33uF	C-EUC0805	C0805
C29	0.1uF	C-EUC0805	C0805

C30	0.1uF	C-EUC0805	C0805
C31	0.1uF	C-EUC0805	C0805
C32	100n	C-EUC0805	C0805
C33	0.1uF	C-EUC0805	C0805
C34	0.1uF	C-EUC0805	C0805
C35	0.1uF	C-EUC0805	C0805
C36	100n	C-EUC0805	C0805
C37	100n	C-EUC0805	C0805
C38	100n	C-EUC0805	C0805
C39	100n	C-EUC0805	C0805
C40	100n	C-EUC0805	C0805
C41	0.33uF	C-EUC0805	C0805
C42	0.1uF	C-EUC0805	C0805
C43	100n	C-EUC0805	C0805
C44	100n	C-EUC0805	C0805
C45	100n	C-EUC0805	C0805
C46	100n	C-EUC0805	C0805
C47	100n	C-EUC0805	C0805
C48	100n	C-EUC0805	C0805
D1	LED_0805	LEDCHIPLED_0805	CHIPLED_0805
D2	LED_0805	LEDCHIPLED_0805	CHIPLED_0805
D3	BYT08P	BYT08P	TO220ACS
D4	BYT08P	BYT08P	TO220ACS
IC1	AT90CAN128	AT90CAN128	TQFP64
IC2	NT-15	NT-15	NT-15
IC3	NT-15	NT-15	NT-15
J1	J-10	J-10	JUMPER
J2	J-10	J-10	JUMPER
J3	J-10	J-10	JUMPER
JP1	AVR-JTAG-10	AVR-JTAG-10	AVR-JTAG-10
L1	10uH	L-EUL2012C	L2012C
Q1	IRFP253	IRFP253	TO-247AD-V

Q2	IRF740	IRF740	TO220BV
Q3	IRF740	IRF740	TO220BV
Q4	16MHz	CRYTALHC49UP	HC49UP
R1	5k6	R-EU_R0805	R0805
R2	500	R-EU_R0805	R0805
R3	500	R-EU_R0805	R0805
R4	500	R-EU_R0805	R0805
R5	500	R-EU_R0805	R0805
R6	500	R-EU_R0805	R0805
R7	500	R-EU_R0805	R0805
R8	500	R-EU_R0805	R0805
R9	500	R-EU_R0805	R0805
R10	500	R-EU_R0805	R0805
R11	500	R-EU_R0805	R0805
R12	500	R-EU_R0805	R0805
R13	500	R-EU_R0805	R0805
R14	500	R-EU_R0805	R0805
R15	500	R-EU_R0805	R0805
R16	500	R-EU_R0805	R0805
R17	RHS25	HS25	resistor-power
R18	500	R-EU_R0805	R0805
R19	500	R-EU_R0805	R0805
R20	500	R-EU_R0805	R0805
R21	500	R-EU_R0805	R0805
R22	500	R-EU_R0805	R0805
R23	R-EU_R0805	R0805	rcl
R24	R-EU_R0805	R0805	rcl
R25	R-EU_R0805	R0805	rcl
R26	R-EU_R0805	R0805	rcl
R27	R-EU_R0805	R0805	rcl
R28	R-EU_R0805	R0805	rcl
R29	R-EU_R0805	R0805	rcl

R30	R-EU_R0805	R0805	rcl
R31	500	R-EU_R0805	R0805
R32	500	R-EU_R0805	R0805
R33	500	R-EU_R0805	R0805
R34	500	R-EU_R0805	R0805
R35	10k	R-TRIMM3296W	RTRIM3296W
R36	500	R-EU_R0805	R0805
R37	500	R-EU_R0805	R0805
R38	500	R-EU_R0805	R0805
R39	500	R-EU_R0805	R0805
R40	R-TRIMM3296W	RTRIM3296W	rcl
R41	R-TRIMM3296W	RTRIM3296W	rcl
R42	R-TRIMM3296W	RTRIM3296W	rcl
R43	500	R-EU_R0805	R0805
R44	500	R-EU_R0805	R0805
R45	500	R-EU_R0805	R0805
R46	R-EU_R1210	R1210	rcl
R47	500	R-EU_R0805	R0805
R48	500	R-EU_R0805	R0805
R49	500	R-EU_R0805	R0805
R50	500	R-EU_R0805	R0805
R51	500	R-EU_R0805	R0805
R52	500	R-EU_R0805	R0805
R53	500	R-EU_R0805	R0805
R54	500	R-EU_R0805	R0805
R55	500	R-EU_R0805	R0805
S1	DS01	DS-01	switch-dil
SV1	ML20	ML20	con-harting-ml
SV2	ML6	ML6	con-harting-ml
SV3	ML14	ML14	con-harting-ml
TP6	TPPAD1-13Y	TPPAD1-13Y	P1-13Y
TP7	TPPAD1-13Y	TPPAD1-13Y	P1-13Y

TP8	TPPAD1-13Y	TPPAD1-13Y	P1-13Y
TP9	TPPAD1-13Y	TPPAD1-13Y	P1-13Y
TP11	TPPAD1-13Y	TPPAD1-13Y	P1-13Y
TP12	TPPAD1-13Y	TPPAD1-13Y	P1-13Y
U$1	SI8233	SI8233	SO-16DW
U$2	TASTER	TASTER	TASTER
U$3	DIODE	DIODE	TO247
U$4	TASTER	TASTER	TASTER
U$5	TASTER	TASTER	TASTER
U$6	NT-5	NT-5	NT-5
U$7	J-10	J-10	JUMPER
U$8	J-10	J-10	JUMPER
U$9	J-10	J-10	JUMPER
U$10	J-10	J-10	JUMPER
U$11	J-10	J-10	JUMPER
U$12	J-10	J-10	JUMPER
U$13	SI8233	SI8233	SO-16DW
U1	ADS8320E/250	ADS8320E/250	SOP65P490X110-8N
U2	ADS8320E/250	ADS8320E/250	SOP65P490X110-8N
U3	ADS8320E/250	ADS8320E/250	SOP65P490X110-8N
U5	ADS8320E/250	ADS8320E/250	SOP65P490X110-8N
U6	ADS8320E/250	ADS8320E/250	SOP65P490X110-8N
V1	TEL5-1222	TEL5-1222	TEL5W
V2	TMH1205S	TMH1205S	TMH12
V3	TME1215S	TME1215S	TME
V4	TME1215S	TME1215S	TME
V5	TME1215S	TME1215S	TME
V6	TMH1205S	TMH1205S	TMH12
X1	MKDSN1,5/2-5,08	MKDSN1,5/2-5,08	MKDSN1,5/2-5,08
X2	MKDSN1,5/2-5,08	MKDSN1,5/2-5,08	MKDSN1,5/2-5,08
X3	MKDSN1,5/2-5,08	MKDSN1,5/2-5,08	MKDSN1,5/2-5,08
X4	MKDSN1,5/2-5,08	MKDSN1,5/2-5,08	MKDSN1,5/2-5,08

X7	MKDSN1,5/2-5,08	MKDSN1,5/2-5,08	MKDSN1,5/2-5,08
X8	MKDSN1,5/2-5,08	MKDSN1,5/2-5,08	MKDSN1,5/2-5,08

A4. Transfer Functions Analysis of Current Sensor

This section presents the corresponding MATLAB code used to analyze the transfer functions of current sensor and ADC.

Obtain transfer function of the current sensor from its datasheet:

```
s=tf('s');
a = 1/0.84e-6;
k=a*2.5/15;
c = k/(s+a);
step(c)
figure
bode(c)
Sensor = tf(1.984126984126984e+05,[1
1.190476190476191e+06])*(15/2.5)
figure
[d1 t1] = step(Sensor,35e-6);
hold all
figure
bode(Sensor)
hold all
```

Obtain transfer function of the low pass filter of the ADC:

```
Filter = tf(1,[510*10e-9 1])
figure
step(Filter)
figure
bode(Filter)
```

Delay time factors in ADC, PWM and microcontroller:

```
d_ADC = 4e-6;     % ADC
d_PWM = 16.5e-6;  % PWM
d_uc = 100/16e6;  % Microcontroller
```

Find the total transfer function of the digital controller with the corresponding delays:

```
% Sensor(T.F.) ---> Filter(T.F.) ---> ADC(Delay) --->
Microcontroller(Delay) ---> PWM(Delay)

% Filter(T.F.) ---> ADC(Delay)
G1 = tf(1,[510*10e-9 1],'InputDelay',d_ADC)
G2 = tf(1,[0 1],'InputDelay',d_uc)
G3 = tf(1,[0 1],'InputDelay',d_PWM)

G = Sensor*G1*G2*G3

figure
[d2 t2]= step(Sensor*G1,35e-6);
figure(44)
bode(Sensor*G1)

figure
plot(t1,d1,'LineWidth',2)
hold all
plot(t2,d2,'--','LineWidth',2)
xlabel('time (\mus)')
ylabel('amplitude (-)')
ylim([0 1.1])
grid on
legend('currnet sensor only','current sensor with LPF
and ADC')
```

Printed in the United States
By Bookmasters